Manual of Experimental Ichthyology

The Authors

Dr. S.K. Gahlawat is an Assistant Professor in the Department of Zoology and Aquaculture, CCS Haryana Agricultural University, Hisar. He did his M.Sc. (1989), M. Phil (1990), Ph.D. (1992) in Zoology from Kurukshetra University and Post-doctoral research in fish biotechnology/ immunology from FRS Marine Laboratory, Aberdeen, UK (2004).

Dr. R.K. Gupta is an Associate Professor in the Department of Zoology and Aquaculture, CCS Haryana Agricultural University, Hisar. He did his M.Sc. (1988) and Ph.D. (1992) in Zoology both from, CCS Haryana Agricultural University, Hisar. He has about 100 publications to his credit in various journals of national and international reputes.

Dr. N.K. Yadava is working as Professor in the Department of Zoology and Aquaculture, CCS Haryana Agricultural University, Hisar. He did his M.Sc. (1973) in Zoology from Gujarat University, Ahmedabad with a specialization in fisheries and Ph.D. (1987) in Zoology from CCS Haryana Agricultural University, Hisar. He has published about 100 research papers including semi technical articles in various national and international journals of repute.

Dr. K.L. Jain did his Ph.D. as ICAR Senior Research Fellow in 1985 in the field of Insect Gustatory Physiology. He worked as Head of the Department of Zoology and Aquaculture in CCS Haryana Agricultural University. Presently he is working as Professor in the Department of Zoology and Aquaculture.

Dr. R.C. Sihag worked as Head of the Department of Zoology and Aquaculture in CCS Haryana Agricultural University, Hisar for three years on a tenure post. Presently he is working as Dean, College of Basic Sciences and Humanities, CCSHAU, Hisar.

Dr. V.P. Sabhlok is working as Professor in the Department of Zoology and Aquaculture, CCS Haryana Agricultural University, Hisar. He has published about 50 research articles in various journals and books.

Manual of Experimental Ichthyology

by
S.K. Gahlawat & R.K. Gupta
Associate Professor
N.K. Yadava, K.L. Jain, R.C. Sihag & V.P. Sabhlok
Professor

Department of Zoology and Aquaculture,
CCS Haryana Agricultural University, Hisar – 125 004, India

2024

Daya Publishing House®
A Division of
Astral International Pvt. Ltd.
New Delhi - 110 002

ISBN 97-881-7035-446-8 (HB)
First Published, 2007
Reprinted 2024

Published by : **Daya Publishing House®**
A Division of
Astral International Pvt. Ltd.
– ISO 9001:2015 Certified Company –
4736/23, Ansari Road, Darya Ganj
New Delhi-110 002
Ph. 011-43549197, 23278134
E-mail: info@astralint.com
Website: www.astralint.com

Laser Typesetting : **Classic Computer Services**
Delhi - 110 035

Printed at : **Replika Press Pvt. ltd.**

Foreward

India has made substantial advancement in fisheries sector by taking the advantage of physical and climatic conditions. More than 3000 crore rupees worth of fish and fish products are being exported to other countries generating a handsome amount of foreign exchange. There is a big space for profitable utilization of our great national resources in aquaculture sector. On the other hand consumption of fish and fish products have also increased many folds during the last decade. As their demand has been increased due to the change of food habits of Indians, therefore, there is a wide scope for effective utilization of available water resources for aquaculture by providing basic know how and various techniques used in fish culture.

The manual entitled *"Manual of Experimental Ichthyology"* prepared by our faculty is highly useful which may serve the long felt needs of students, teachers and research scholars. I believe that this work will be of great use for fisheries under-graduate and post-graduate students. In all, it contains four useful units comprehensively covering almost all aspects of fish culture. I congratulate our faculty for their mammoth efforts in bringing out this manual.

S.S. Dahiya
Dean
Post-graduate Studies
CCS Haryana Agricultural University, Hisar

Preface

This manual entitled *"Manual of Experimental Ichthyology"* has been written in order to compile the information to meet the needs of fisheries student at graduate as well as post-graduate level. The prime objective of the manual is that the student should learn about the identification and description of fishes. Various basic methodologies about fish culture have been described in this manual. This manual comprises of four units in all. The first unit describes some basic aspects of fish culture and their identification methodologies. The second unit comprises of various physiological units which may serve the long felt need of students, teachers and research scientists. The third unit consists of a comprehensive range of basic experiments used in water analysis used in fish culture. In the last, fourth unit consists of various aspects of fish diseases. This unit can be useful to control the evergrowing menace of disease faced by aquaculture industry.

The authors are thankful to the CCS Haryana Agricultural Universty authorities and the publisher for encouraging us to write this manual and to provide financial assistance for this manual. We are also indebted to various authors whose valuable work have been cited in this manual.

Authors

Contents

Unit–I

Exercise No. 1

Definition, General Classification and Measurement of External Parts of the Body of a Fish

Definition of Fish

Fishes are cold blooded aquatic vertebrates which breathe by means of pharyngeal gills, propelling and balancing themselves by the means of fins. Some 20,000 species are known to inhabit water bodies of the various descriptions.

General Classifications of Fishes

The living forms included in superclass Pisces are divided into four classes, *viz.* Elasmobranchii, Holocephali, Dipnoi and Teleostomi, of which, Dipnoi does not occur in India. The diagnostic characteristics of Indian representatives of the three classes are:

Class-Elasmobranchii

Skeleton cartilagenous; skin of body covered with placoid scales or naked; 5-7 gill openings on each side; no gill cover, teeth distinct; cloaca present; males with claspers, *e.g.* Rays and Sting Rays

Class-Holocephali

Skeleton cartilagenous, skin of the body covered with placoid scales; 5 gill arches; 4 gill openings protected by a gill cover or the operculum; teeth united to form grinding plates; no cloaca; males with claspers, *e.g.* Chimarea.

Class-Dipnoi

The Dipnoi (double breathers) is an important group of bony fishes. They are also called as "Lung fishes" and represent transitional group between fishes and amphibians. Class-Dipnoi comprises only two orders: Monopneumona (with single lung), and Dipneumona (with double lungs). The fishes of this class are having long fusiform body, strong heterocercal tail, archipterygial types paired fins, two dorsal fins and large heavy cosmoid scale, ossified jaws and dentitures highly specialized with suppressed marginal teeth in both the jaws. These fishes frequently come to surface and breathe air.

Class-Teleostomi

Skeleton bony; skin of the body covered with ganoid, catenoid scales or naked; 5 gill arches; 4 gills; a pair of lateral gill-opening confluent as two lateral slits with operaculum; usually no cloaca. Almost all freshwater fishes are covered under this class.

Different Parts of the Body of a Fish (Figure 1)

1. Eye (without eye lids)
2. External nostrils

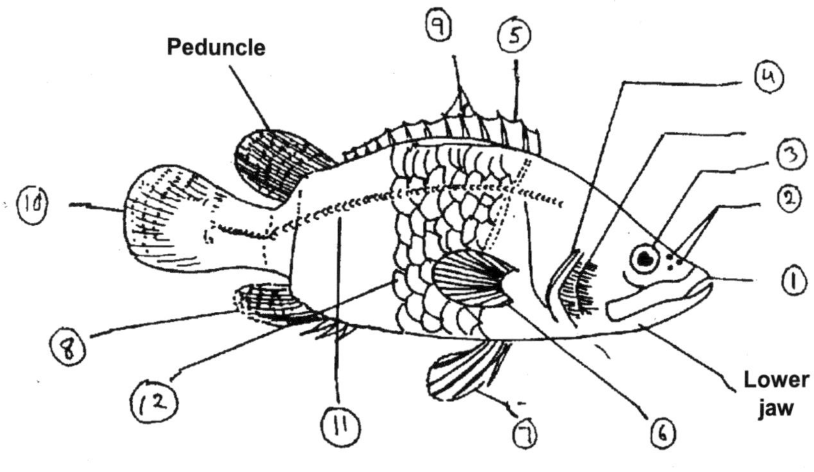

Figure 1: Body Parts of a Fish

(1) Mouth; (2) External nostrils; (3) Eye; (4) Operculum; (5) Dorsal fin; (6) Pectoral fin; (7) Ventral fin; (8) Anal fin; (9) Fin rays; (10) Caudal fin; (11) Lateral line; (12) Scale

3. Mouth consists of Pre-maxilla Maxilla, Mandibles, Teeth Grinding plates, Lip, Mouth cavity, Mouth tentacles
4. Operculum consists of pre-operculum, inter-operculum and sub-operculum
5. Fins are Dorsal, Ventral, Pectoral, Anal, Caudal
6. Fin ray spines (Dorsal, Pectoral, Anal, Ventral)
7. Additional Dorsal Fin or Peduncle
8. Scales (Placoid, Ganoid, Cycloid and Ctenoid)
9. Lateral line (LI)
10. Latero-transversal line (Ltr).

Measurement of Different Parts of the Body of a Fish

Name of the Organ	*Way of measurement*
Eye	By taking total diameter of eye
Snout	By taking length between mouth and eye
Length of head	By measuring distance between mouth and postoperculum
Length of body	By measuring distance between mouth and commencement of caudal fin
Total length	By measuring vertical distance between mouth to end of caudal fin
Depth of body	By measuring vertical distance between peduncle (dorsal fin) and anal fin

Fish Formula (FF)

It may be defined as the formula by which the exact number and measurements of different parts of the body/number of fin rays of different fins and number of scales present on the lateral line as well as on the latero-transverse line may be counted. It is the key to identify and differentiate among species of the same genera, *e.g.*

$$FF = D14\text{-}16, A8, V_{9}, P_{21}, C_{19}, LI_{40\text{-}42}, Ltr_{7.5\text{-}9}$$

where,

D = No. of fin-rays of Dorsal fin,

A = No. of fin-rays of Anal fin,

V = No. of fin-rays of Ventral fin,

P = No. of fin-rays of Pectoral fin,

C = No. of fin-rays of Caudal fin,

LI = No. of scales present on the lateral line of the body of fish

Ltr = No. of scales present on the latero-transverse line of the body of fish.

Exercise No. 2

To Study the Habit and Habitat of Some Important Culturable Freshwater Fishes

The class Teleostomi commonly known as teleosts includes families Cyprinidae, Siluridae and Ophiocephalidae. The general description of these families is as follows:

Family: Cyprinidae

- ☆ Body oblong or elongated, abdomen usually rounded, but may be compressed, margin of upper jaw is formed by the pre-maxillaries
- ☆ Operculum is in four pieces. Mouth toothless, but sometimes form one to three rows of teeth in the interior pharyngeal bones
- ☆ A single rayed dorsal fin
- ☆ Head scaleless. Body scaled but rarely scaleless.
- ☆ Mostly herbivorous in food habits.

Family: Siluridae

- ☆ Margin of the upper jaw formed mainly by pre-maxillaries, Maxilla is rudimentary and often constitutes the base of a barbel
- ☆ Barbels are present
- ☆ No sub-opercle

- ☆ Either rayed dorsal fin or adipose dorsal fin, may be present or absent
- ☆ Skin scaleless
- ☆ Mostly carnivorous in food habits
- ☆ Mouth is armed with teeth.

Family: Ophiocephalidae

- ☆ Body elongated, sub-cylindrical anteriorly, head depressed having superiorly somewhat plate like scales
- ☆ Eyes are lateral. Gill opening wide
- ☆ Teeth are present in the jaws or vomer or plates, some of which may be conical
- ☆ A single, long, spineless dorsal fin
- ☆ Ventral fin, thoracic fin may be present or absent, when present, consisting of six rays
- ☆ Scales are of large, moderate or small in size
- ☆ Carnivorous in food habits.

Identification of Some Economically Important Cultivable Fishes

1. Family: Cyprinidae

(a) Major Carps

(*i*) *Catla catla*

(*ii*) *Labeo rohita*

(*iii*) *Cirrhina mrigala*

(b) Minor Carps

(*i*) *Labeo calbasu*

(*ii*) *Cirrhina reba*

(*iii*) *Barbus sarana*

(c) Exotic Carps

(*i*) *Ctenophryngodon idella*

(*ii*) *Hypopthalmicthys molitrix*

(d) Common Carp

(*i*) *Cyprinus carpio* var. communis

(*ii*) *Cyprinus carpio* var. specularis

(*iii*) *Cyprinus carpio* var. nudus

2. Family: Siluridae

(a) Cat Fishes

(*i*) *Wallago attu*

(*ii*) *Hateropnuestes fossilis*

(*iii*) *Clarias magur*

(*iv*) *Mystus seenghala*

3. Family: Ophiocephalidae

(b) Murrel Fishes

(*i*) *Ophiocephalus striatus*

(*ii*) *Ophiocephalus punctatus*

(*iii*) *Ophiocephalus marulius*

(*iv*) *Ophiocephalus gachua*

Indian Major Carps

(a) Catla *(Catla catla)*

Class	:	Teleostomi
Order	:	Physostomi
Family	:	Cyprinidae
Genus	:	*Catla*
Species	:	*catla*

☆ It is commonly called catla and is distributed in almost all States of India, though, its natural distribution extends only to the Godavari River in Andhra Pradesh, but now distributed in almost all the river systems.

☆ It is the fastest growing carp in India.

☆ Characteristically, bulged body with a conspicuous head, large upturned mouth, non fringed lips, devoid of barbells and a dorsal fin with 14 to 16 branch rays, distinguish the adult fish and advanced fingerlings.

- ☆ The body is ordinarily dull, silver white, but tends to be rather darkish in weedy waters.
- ☆ The fry, from half to one inch in size, are distinguished by the relatively large head with reddish gills.
- ☆ Catla is a surface and column feeder and its upturned mouth and large gill rakers are adapted to feed on numerous organisms, floating in water.
- ☆ Submerged weeds are of no use to catla and in ponds with an overgrowth of weeds, its growth is generally unsatisfactory.
- ☆ It is reported to grow very quick even unto three to four inches per month.
- ☆ In normal stocked waters, a growth of 15 to 18 inches in the first year can usually be expected.
- ☆ In lifetime, catla grows to over four feet in length.
- ☆ Fish in the second year, are ordinarily sexually ripe. Specimens 22 months old measuring 18 inches in total length and weighing about 3 pounds, have been observed sexually mature.
- ☆ Catla breeds in river during monsoon months from June to August.
- ☆ Usually it does not breed in pond or lakes, especially in captivity.

FF=D.14-16, P.21, V.9, A.8 (3/8), C.19, L1. 40-42, Ltr. 7½/9

(b) Rohu (*Labeo rohita*)

Class	:	Teleostomi
Order	:	Physostomi
Family	:	Cyprinidae
Genus	:	*Labeo*
Species	:	*rohita*

- ☆ A quick growing carp, with almost the same distribution as catla and considered a tastier fish than any other Indian carp.
- ☆ It is commonly known as rohu, distributed to Bengal, Bihar, Assam, Uttar Pradesh, Punjab, Orissa and Andhra Pradesh.

- ☆ Rohu is easily distinguished by its small or pointed head, almost terminal mouth with fringed lower lip, dull reddish scales on the sides and pink reddish fins.
- ☆ The dorsal fin has 12 to 13 branched rays.
- ☆ The body is more linear than that of catla.
- ☆ Rohu is a column-bottom feeder and its rather anterior fringe-lipped mouth is well adapted to browsing in the shallow pond's bottom also. Considerable quantities of bottom sand or mud, vegetable debris or decaying leaves of aquatic plants, planktonic algae, etc. constitute the stomach contents.
- ☆ Water fleas or micro animalcules are of no use as food.
- ☆ Rohu grows very quick, though perhaps, relatively a little slower than catla.
- ☆ A growth of 14 to 16 inches can normally be expected in the first year in a well stocked fish pond. Sexual maturity is attained towards the end of the second year.
- ☆ Rohu grows to over three feet in length.
- ☆ Like catla, rohu also breeds in rivers and also in the bund type tanks, during the monsoon and its frys are collected from these habitats for rearing and stocking.

FF=D.15-16, P.17.V.9.A.7(2/5), C.19. LI.40-42, Ltr. 7½/9

(c) Mrigal (*Cirrhiana mrigala*)

Class	:	Teleostomi
Order	:	Physostomi
Family	:	Cyprindae
Genus	:	*Cirrhina*
Species	:	*mrigala*

- ☆ Next in importance to catla and rohu for culture purposes is the mrigal.
- ☆ Common in major river systems including river Godavari.
- ☆ Locally called mori, distributed mainly in states of Bihar, Bengal, Orissa, Uttar Pradesh, Punjab and Andhra Pradesh.

- ☆ It is easily distinguished by the relatively linear body, small head with rather blunt snout, terminal mouth with thin non-fringed lips, bright silvery body and reddish fins.
- ☆ The dorsal fin has 12-13 branched rays. The young ones can easily be distinguished by their blunt snout and thin lips.
- ☆ In slight larger specimens, the scales at the side of the body have their edges pigmented dark.
- ☆ The fins also become reddish.
- ☆ Mrigal is a bottom-column feeder.
- ☆ The fingerlings and adults take relatively large quantities of decaying vegetable and organic debris, phyto-planktons, sand and mud, as compared to rohu.
- ☆ The proportion of animal matter in the diet is generally poor.
- ☆ The thin terminal lips are adapted to picking up things from the mud and form the invariable presence of appreciable quantities of sand and mud in the gut, algae in the fresh or decaying conditions are also picked up from the bottom.
- ☆ Mrigal grows slower than catla and rohu.
- ☆ The species attains a length of over three feet.
- ☆ Mrigal becomes sexually mature only when about two years old.
- ☆ The breeding season and source of fry are the same as those of catla and rohu.

FF=D.15-16, P.15, V.9, A.8 (3/5), C.15, L1.40-45, Ltr. 6½-7/8½

Minor Carps (Family: Cyprinidae) (Figure 2)

(a) Kalbasu (*Labeo calbasu*)

Class : Teleostomi

Order : Physostomi

Family : Cyprinidae

Genus : *Labeo*

Species : *calbasu*

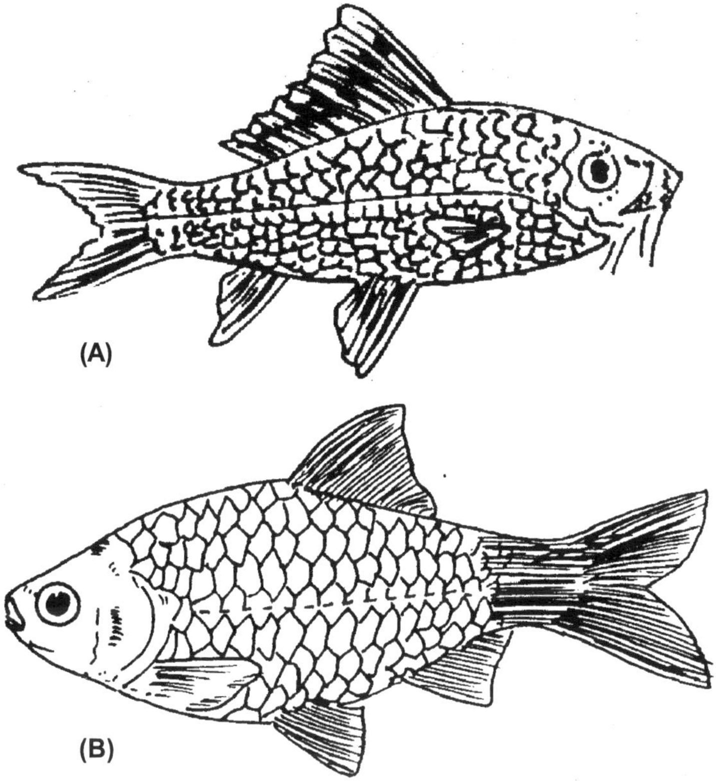

Figure 2: Indian Minor Carps
(A) *Labeo calbasu*; (B) *Barbus sarnana*

- ☆ Kalbasu occurs almost throughout India, except in Malabar.
- ☆ It does not generally occur in large numbers, but whenever available, it is stocked in ponds.
- ☆ Though attaining a maximum length of 2.5 feet or more, it is relatively slow growing fish.
- ☆ The conspicuous, uniformly darker (blackish gray) colour of the body, the small tapering head with the sub-terminal fringed-lipped mouth, and conspicuous barbels are its distinguishing characters, the dorsal fin has 12 to 13 branched rays as in rohu and mrigal.
- ☆ The fry are easily distinguished by their peculiar pigmentation.

- ☆ Kalbasu feeds on snails and worms at the bottom of the pond and also on algae and other decayed matters, but the animal portion of the diet is generally very small.
- ☆ In well stocked ponds, an average annual growth of 10-12 inches may be expected in the first year. Often small specimens, about 8.5 to 9 inches long, are sexually ripe; there is no direct evidence that they attain maturity in the first year. Usually in the large specimens, maturity is attained in the second year only.

FF=D.16-18 P.19, V.9, A.7(2/5), C.19, LI.40-44, Ltr. 7½/8

(b) Reba (*Cirrhina reba*)

Class	:	Teleostomi
Order	:	Physostomi
Family	:	Cyprinidae
Genus	:	*Cirrhina*
Species	:	*reba*

- ☆ It is a widely distributed minor carp, but has not so far been recorded in Malabar.
- ☆ Though, its maximum recorded size is 12 inches, it is extensively used for stocking purpose in Madras.
- ☆ The characteristics hexagonal scales on the body, the lateral dark pigment band on either side and the small tapering head are its distinguishing features.
- ☆ The dorsal fin has only eight to nine branched rays.
- ☆ A dark pigmented band, originating on the snout and extending up to the tail on either side, is also a distinguishing character.
- ☆ Reba appears to subsist largely on planktonic algae which form about sixty per cent of the food.
- ☆ Considerable quantities of mud, sand and debris are also met within the gut contents.
- ☆ The young ones feed voraciously on zooplanktons and sometimes grow faster than even the young ones of catla.
- ☆ Reba has more pronounced plankton feeding habits.

- ☆ Though a minor carp, *Cirrhina reba* grows fairly quickly, and by the end of first year of life, attains a length of about 11 inches.
- ☆ It breeds in rivers during monsoon months. Carp fry are collected from reverine habitats.

FF=D.10-11, P.16, V.9, A.8(3/5), C.19, LI. 35-48, Ltr. 7/9

(c) Sarana (*Barbus sarana*)

Class : Teleostomi

Order : Physostomi

Family : Cyprinidae

Genus : *Barbus*

Species : *sarana*

- ☆ This is medium sized fish and attains a length of over a foot and distributed throughout India.
- ☆ Though, there is no regular culture of this species in certain localities, particularly in Chennai (Madras), fingerling are collected for stocking in ponds.
- ☆ It is a voracious feeder, consuming quantities of decaying macro-vegetations, gastropod shells, worms, insects and filamentous algae.
- ☆ It breeds in rivers during monsoon months but not in ordinary ponds.
- ☆ The young ones are caught along with the fry of major carps and stocked in ponds.
- ☆ Usually only a few specimens survive and grow rapidly, attaining a length of about 10 inches in the first 10 months of their life, and sexual maturity also at the end of this period.

FF=D.11, P.15, V.9, a.8 (3/5), C.19, L1. 32-34, Ltr. 5½ -6/6

Cat Fishes (Family: Siluridae) (Figure 3)

(a) Freshwater Shark (*Wallago attu*)

Class : Teleostomi

Order : Physostomi

Family : Siluridae

Genus : *Wallago*

Species : *attu*

☆ One of the largest of the freshwater catfishes attaining a length of about six feet, *W. attu* is widely distributed throughout India.

☆ It is highly predatory in nature and highly destructive to carps. Its large mouth, well armed with teeth, enables it to successfully attack and destroy even medium sized carps.

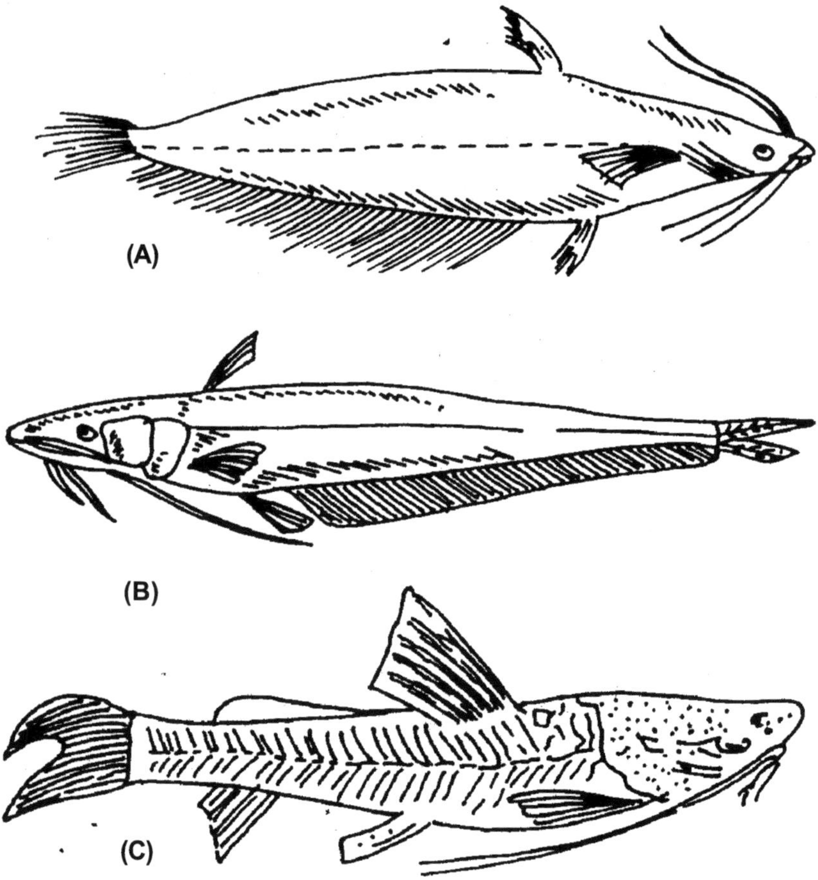

Figure 3: Cat FIshes (Family: Siluridae)
(A) *Heteropnuestes fossilis* (Singhi);
(B) *Wallago attu* (Freshwater Shark);
(C) *Clarias magur* (Magur)

- ☆ A variety of items including limbs have been observed in its stomach.
- ☆ It breeds in rivers during the rainy months.
- ☆ It is reported to attain a length of about 30 inches in the first year of its life.
- ☆ On account of its predacious habits, this species should be carefully removed from carp ponds.

(b) Singhi (*Heteropnuestes fossilis*)

Class : Teleostomi
Order : Physostomi
Family : Siluridae
Genus : *Heteropnuestes*
Species : *fossilis*

- ☆ This is a common catfish with accessory respiratory organs and therefore, able to breath atmospheric oxygen.
- ☆ It is available throughout India and attains a length of over a foot.
- ☆ Though not as highly price food fish as Murrel, its flesh is supposed to have invigorating qualities.
- ☆ The fish is very much feared for its poisonous pectoral spines which can inflict very painful wounds leading to local inflammation and even fever.
- ☆ It is predacious in nature but not markedly piscivorous. Insects, ostracodes, worms, algal matter, organic debris and fish and fish remains constitute the main items of food.
- ☆ Varying quantities of mud or sand are also often taken in. It breeds during April-June but very little is known of its history/growth.

(c) Magur (*Clarias magur*)

Class : Teleostomi
Order : Physostomi
Family : Siluridae
Genus : *Clarias*
Species : *magur*

- ☆ This is a catfish with habits very similar to that of singhi but not so common as the latter, though found throughout India and growing to much larger size.
- ☆ Being an esteemed food fish, it is always in high demand.
- ☆ Magur is predatory like singhi but not so markedly piscivorous.
- ☆ Medium sized specimens have been found to take prawns in appreciable numbers, aquatic insects and their larvae, sand and mud occasionally mixed with algal filaments.
- ☆ The fish breeds in ponds during April-June but in different types of environments.

(d) Seenghara (*Mystus seenghala*)

Class	:	Teleostomi
Order	:	Physostomi
Family	:	Siluridae
Genus	:	*Mystus*
Species	:	*seenghala*

- ☆ This is another large reverine catfish, distributed almost throughout India.
- ☆ It attains a length of about six feet and is sill in greater demand than *W. attu* as a food fish.
- ☆ The fish is a predator with marked piscivorous tendencies.
- ☆ *Barbus filamentous* and *Cirrhina reba* have been found most frequently in its stomach in the Cauvery river.
- ☆ Insects and their larvae, crustaceans, gastropods and very rarely aquatic weeds are also consumed.
- ☆ The fish breeds in rivers as well as in ponds during April to July.
- ☆ Its breeding habits are peculiar because it shows parental care in rearing its eggs and young ones.

Murrel or Live Fishes (Family: Ophiocephalidae) (Figure 4)

(a) Common Murrel (*Ophiocephalus striatus*)

Class	:	Teleostomi
Order	:	Physostomi
Family	:	Ophiocephalidae
Genus	:	*Ophiocephalus*
Species	:	*striatus*

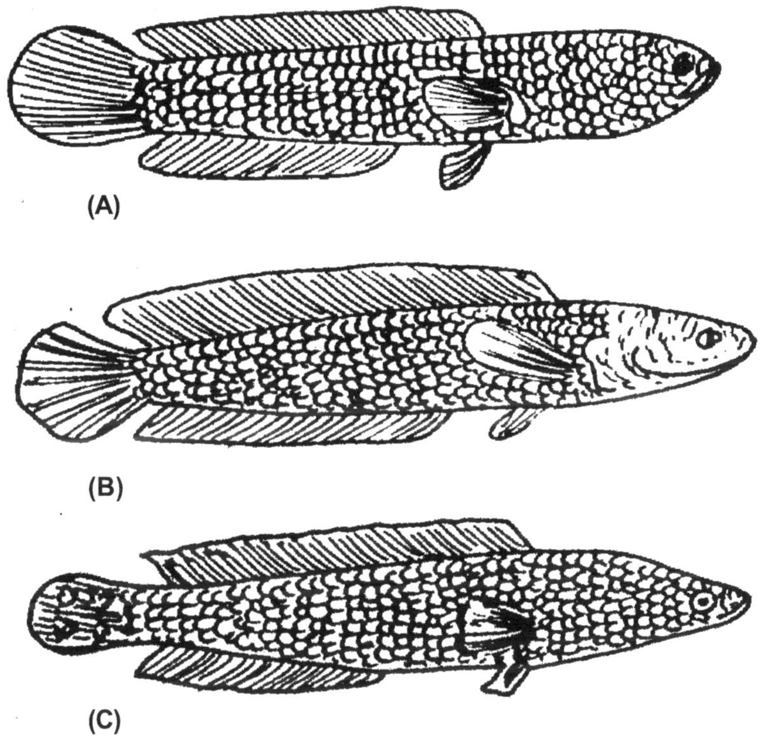

Figure 4: Murrel Fishes (Family: Ophiocephalidae)
(A) *Ophiocephalus punctatus* (Kaveri Murrel); (B) *Ophiocephalus striatus* (Common Murrel); (C) *Ophiocephalus marulius* (Large Murrel)

☆ Murrel forms the bulk and mainstay of the pond fisheries in several parts of India.

☆ It is regarded as an excellent table fish in Punjab, Madhya Pradesh, Hyderabad, Mysore and Travancore–Cochin, and is highly priced and is in greater demand than carps.

☆ *Ophiocephalus striatus* is perhaps the commonest and, economically, the most important food fish.

☆ The different species of murels resemble each other closely.

☆ Being a predator, it feeds on minnows, minor carps, major carps, fingerlings, insects, frogs and other live food, fries feed on zooplanktons and fingerlings feed mainly on insect larvae and young fish, if available.

☆ It breeds in ponds almost throughout the year, with a peak period immediately preceding, and during the monsoon months.

☆ Sexual maturity is attained at a length of about 9.5 or 10 inches.

☆ Eggs are laid in floating mats or scums and parental care is also observed in this fish.

☆ Under favourable conditions, a quick growth of up to 19 inches has also been recorded.

(b) Kaveri Murrel (*Ophiocephalus punctatus*)

Class : Teleostomi

Order : Physostomi

Family : Ophiocephalidae

Genus : *Ophiocephalus*

Species : *punctatus*

☆ This is one of the small species of the genus attaining a foot in length and subsiding largely on small fishes and aquatic insects.

☆ Breeding season and habits are similar to those of *O. striatus.*

☆ Very little is known about its growth, but sexual maturity is probably attained in the first year of its life as in *O. striatus.*

(c) Large Murrel *(Ophiocephalus marulius)*

Class : Teleostomi

Order : Physostomi

Family : Ophiocephalidae

Genus : *Ophiocephalus*

Species : *marulius*

- ☆ Probably the largest among the murrels, *O. marulius* attains a length of about four feet and is found almost throughout India.
- ☆ It prefers large bodies of water like lakes, reservoirs and swamps, though it is also found in rivers.
- ☆ It does not breed in small ponds. It is so highly predacious, even cannibalistic, that when stocked in confined waters like irrigation wells, the survival is limited to only a very few.
- ☆ It breeds from April to June and breeding habits are similar to those of *O. striatus.*
- ☆ The young ones grow fast, a maximum growth of about 30 inches at the end of first year of its life is on record.

(d) Small Murrel *(Ophiocephalus gachua)*

Class : Teleostomi

Order : Ophiocephalidae

Family : Ophiocephalus

Genus : *Ophiocephalus*

Species : *gachua*

- ☆ This is called a minor or small murrel because of its small size, *O. gachua* is of very limited economic significance, it attains a maximum size of four to six inches, feeds on small fries, fingerlings, aquatic insects, micro crustaceans and other small aquatic insects.
- ☆ Sexual maturity is attained at the end of first year.

Exotic Carps (Family: Cyprinidae) (Figure 5)

Some exotic fishes of sport, food and public health significance have been introduced in India. As these fishes were very hardy so they easily flourished in this country. Following are more important from the point of view of fish culture.

(a) Grass Carp (*Ctenopharyngodon idella*)

Class	:	Teleostomi
Order	:	Physostomi
Family	:	Cyprinidae
Genus	:	*Ctenopharyngodon*
Species	:	*idella*

- ☆ This fish is the natural inhabitant of the flat land rivers of China and the middle and lower sections of river Amur in USSR.
- ☆ Grass carp has been introduced in many countries including India from Hong Kong.

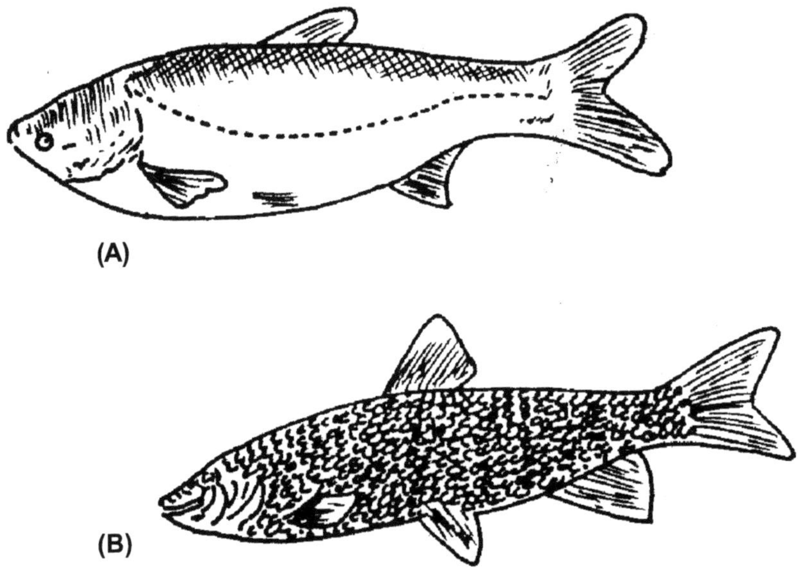

(A)

(B)

Figure 5: Exotic Carps (Family: Cyprinidae)
(A) *Hypopthalmicthys molitrix* (Silver Carp);
(B) *Ctenopharyngodon idella* (Grass Carp)

☆ Head is broad with short round snout, upper jaw slightly longer than the lower, no barbel, mouth sub-terminal and fin round, colour of the body dark grey above and silvery on the belly, all fins dark, base of each scale dark brown.

☆ The maximum length reported is 120 cm and weight up to 32 kg.

☆ Grass carp, though a freshwater fish, is also able to tolerate slightly brackish water like the silver carp and is almost exclusively a vegetarian.

☆ Like other Cyprinids it has a toothless mouth but has strong and specialized pharyngeal teeth for grasping aquatic vegetations.

☆ These teeth occur in two rows.

☆ The spawning season of grass carp is in summer approaching to monsoon to monsoon season in the flooded rivers.

(b) Silver Carp (*Hypopthalmicthys molitrix*)

Class : Teleostomi

Order : Physostomi

Family : Cyprinidae

Genus : *Hypopthalmicthys*

Species : *molitrix*

☆ Silver carp naturally occurs in the river systems of central and south China and in Amur river of USSR, and now it is cultured in many countries including India.

☆ This was brought to India in 1959 from Japan for culturing at CIFRI, Cuttack and Orissa. In due course, it was bred successfully by hypophysation and the fries were distributed from Cuttack to various States of India.

☆ The diagnostic characters of silver carp are the oblong and slightly compressed body, head pointed, bluntly rounded snout, slightly protruding lower jaw, small scales, abdomen keeled from the throat to the ventral, origin of dorsal fins behind ventral and dark fins.

- ☆ In case of this fish, fry and adult sustain on *Flagellata. Dinoflagellata, Myxophycea, Protozoa* and *Rotifera*, supplemented with decayed macro-vegetations and detritus.
- ☆ When kept in ponds, it thrives well on artificial feeds, such as bone meal, rice bran and flour.
- ☆ The fish attains full maturity by the time they are two years old.
- ☆ This is an esteemed food fish and shows encouraging production when stocked with other indigenous carps.

(c) Common Carp (*Cyprinus carpio*) (Figure 6)

Class	:	Teleostomi
Order	:	Physostomi
Family	:	Cyprinidae
Genus	:	*Cyprinus*
Species	:	*carpio*

- ☆ This is a world wide most extensively cultivated species.
- ☆ Although exotic of India, it is already being cultivated in the country singly as well as along with Indian major carps.
- ☆ The fish is voraciously omnivorous, efficiently converting the food ingested into flesh, grows very fast and prone to artificial feeds.
- ☆ It is non predatory, when cultured along with major carps, it competes for food mostly with mrigal and kalbasu.
- ☆ The strain of common carp was obtained at Cuttack from Bangkok in 1957, is found to procleate prolifically in the plains of India and has been distributed to different States of the Country.

The common carp has three recognized varieties, differentiable according to the pattern of scaling.

(*i*) The Scale Carp (*Cyprinus carpio var.* communis)

Class	:	Teleostomi
Order	:	Physostomi

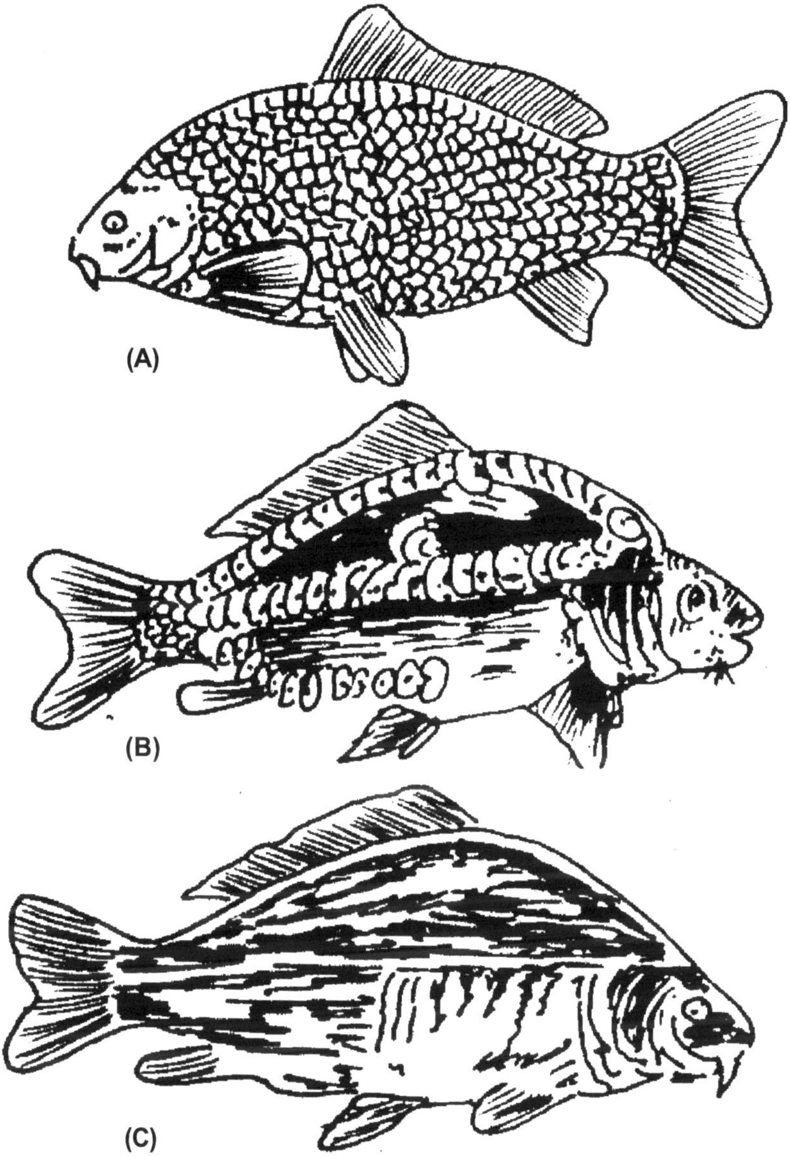

Figure 6: Common Carp (*Cyprinus carpio*)
(A) Scale Carp (*Cyprinus carpio var.* Communis)
(B) Mirror Carp (*Cyprinus carpio* var. specularis)
(C) Leather Carp (*Cyprinus carpio* var. nudus)

Family : Cyprinidae

Genus : *Cyprinus*

Species : *carpio*

Variety : communis

- ☆ It has its body fully covered by regularly arranged rows of scales.
- ☆ This is the exotic form, now extensively cultivated in our country in the various States including Haryana.

(*ii*) The Mirror Carp (*Cyprinus carpio* var. specularis)

Class : Teleostomi

Order : Physostomi

Family : Cyprinidae

Genus : *Cyprinus*

Species : *carpio*

Variety : specularis

- ☆ It has its body covered unevenly with a few large and bright scales.
- ☆ A large area of the body is, however, naked.
- ☆ This species is extremely variable and is also cultivated in the State of Haryana.

(*iii*) The Leather Carp (*Cyprinus carpio* var. nudus)

Class : Teleostomi

Order : Physostomi

Family : Cyprinidae

Genus : *Cyprinus*

Species : *carpio*

Variety : nudus

- ☆ It has its body completely devoid of scales except for a single row of somewhat degenerated scales along the base of the dorsal fin and sometimes extending as far as the tail or even from head to tail.
- ☆ Sometimes tiny scales are also present at the bases of other fins.

Exercise No. 3

Methods of Identification of Fish Seed of Culturable Species (Spawn, Fry, Fingerlings)

Aim and Objectives

To identify the spawn, fry and fingerlings of major carps and other food fishes and differentiate between spawn, fry and fingerlings.

Relevant Information

☆ Fish hatchlings measuring up to 8 mm in length are known as spawn. Fish seed above 8 mm and upto 40 mm in length are known as fry. Fish seed above 40 mm and upto 150 mm in length are known as fingerlings.

☆ A typical fish pond consists of three food niches: surface, column and bottom. Different varieties of fishes feed in different niches of the pond. For example, catla is surface feeder. Rohu is column feeder and mrigal is a bottom feeder. While stocking the ponds with fry and fingerlings different groups of fish with different feeding habits are stocked so that food available in all the three niches is properly utilized for maximum growth of fish seed.

☆ The following are identification characteristics of different species of spawn, dry and fingerlings. Carp spawn show jerky forward movements, whereas, the spawn of catfishes exhibit wriggling movements.

☆ Carp spawn, when it is 5.0 to 5.5 mm long is differentiated into desirable and undesirable varieties. In the desirable carp spawn, the yolk sac is wide anteriorly, the embryonic fin fold is continuous, the eyes pigmented and the mouth is formed. In the undesirable carp spawn, the yolk sac is disappearing, rudimentary or absent. The embryonic fin-fold is discontinuous, there are rudiments of dorsal fin with or without fin rays, the mouth is formed and these are pigmented.

☆ Carp fishes are divided, broadly into (*i*) major carps and (*ii*) minor carps. Major carps grow faster and are bigger in size as compared to minor carps. At the fry stage, the carps can be easily differentiated into major and minor.

Major Carps

The number of undivided dorsal fin rays is more than eleven (*e.g. Catla catla, Labeo rohita, L.finbriatus, L.calbasu, L. gonius, Cirrhinus mrigala*). Individual characteristics of major carps fry of different species are as under:

Catla catla

Large head, dorsal profile convex and ventral profile concave. No distinct spot on the caudal fin, or at the caudal peduncle, margins of the dorsal and caudal fins darker. First ray of the dorsal fin is black. Opercular region brightly reddish. No barbels. Lips thick but not fringed.

Labeo rohita

A dark diffuse transverse dark band present at the caudal peduncle. A pair of whitish or grayish maxillary barbels present, lips fringed.

Labeo calbasu

At the earlier stages, the fry are darker in colour. A semilunar yellow band appears on the base and gradually the body gets the appearance of having alternate yellow and black bands. The basal portions of the fins are black which gradually spread over the fins. Two pairs of thick black barbels (maxillary and rostral) present. Lips fringed, mouth narrow.

Labeo gonius

Body marked with a number of very faint longitudinal white stripes. An indistinct spot present at the caudal preduncle. A pair of maxillary barbels present. Lips fringed, mouth narrow.

Labeo fimbriatus

Head small and mouth terminal. A conspicuous dark caudal spot present, small head with thick fringed lips. A pairs of barbels present. Lips fringed, mouth narrow.

Cirrhinus mrigala

Small head and slender body. A more or less triangular dark spot at the caudal peduncle. No barbels visible. Lips thin, not fringed.Tip of the lower lobe of the caudal fin reddish.

Minor Carps

The number of undivided dorsal fin rays are eleven or less than eleven (*e.g., Labeo bata, Cirrhinus reba and Puntius surana)*

Labeo bata

A bright dark caudal spots more or less semi-circular in shape present at the caudal peduncle. This spot gradually becomes more or less circular in outline. No barbels visible. Lips not fringed.

Cirrhinus reba

Slender body. Presence of three distinct dark spot: two on the caudal fin and one at the base of the caudal peduncle. Pigmented spots on the lateral line, which gradually take the shape of a dark lateral band. Mouth terminal, no barbels, lips not fringed.

Puntius surana

Body deep and moderated compressed. A small black vertical band extending from the point of insertion of the dorsal fin. An oval dark spot present at the caudal peduncle. Lips thin and not fringed. Maxillary barbels present. The first ray of the dorsal fin blackish.

Materials Required

- ☆ Spawn, fry and fingerlings of different species.
- ☆ Formalin
- ☆ Glass bottle with a cork
- ☆ Petri dishes

- ☆ Beakers
- ☆ Enamel trays
- ☆ Plastic buckets
- ☆ Mug
- ☆ Freshwater
- ☆ Pipette
- ☆ Fine brush
- ☆ Forceps
- ☆ Hand lens
- ☆ Hand and drag nets
- ☆ Pencil and eraser
- ☆ A few drawing sheets

Procedure

- ☆ Take a sample of fish seeds in a beaker and observe the movements. (Generally fish seeds showing wriggling movements are catfish seed; and those showing jerk, forward movements are carp seed).
- ☆ Take a Petri dish with formalin and place a sample of fish seed in it.
- ☆ Measure the length and decide whether the sample consists of spawn, fry or fingerlings. If the sample consists of carp seed, count the dorsal fin rays and determine whether they belong to major carps or minor carps.

Study of Individual Characteristics

Make diagram of different species with a pencil on the drawing sheets and note down individual characteristics. Name the individual species by drawing inferences from individual characteristics.

Precautions

- ☆ Handle the spawn, fry and fingerlings carefully so that they are not injured.
- ☆ Stock only healthy fry and fingerlings into the ponds.
- ☆ Measure length of spawn, fry and fingerlings correctly for differentiation into the respective groups for stocking purpose.

Exercise No. 4

Methods of Stocking of Spawn and Fry in Nursery Ponds

Aims and Objectives

To known the importance of stocking of spawn and fry in nursery ponds and to get a high percentage of survival from spawn to fry and fry to fingerlings.

Relevant Information

- ☆ Spawn is a tiny fish (5-6 mm in length) in which the yolk sac is absent and the digestive system is developed and it performs horizontal, free movements in the water. This spawn up to 9 mm size is stocked in the nursery ponds and after it has attained the size of 30-40 mm, it is called fish fry. The fry can be stocked in the rearing ponds to grow up to fingerling size (40-50mm). Nursery ponds are for the stocking of fry.
- ☆ Nursery ponds are the small water bodies in which proper care of feeding and other necessary steps can be undertaken easily. The merits of stocking spawn and fry in nursery and rearing ponds respectively are as under:
  - ☆ Spawn or fry can get the feed easily;
  - ☆ More seed can grow in a given area under proper care;
  - ☆ Spawn and fry grow faster;
  - ☆ It is easy to remove the predators and other harmful organisms from nurseries;

- ☆ Survival percentage from spawn to fry and fry to fingerlings is high.
- ☆ It is easy to harvest the seed.
- ☆ Assessment of seed can be done in nurseries by simple netting.
- ☆ Generally, the spawn is stocked at the rate of 50-60 lakhs/ hectare as a conventional rate of stocking but under modern nursery management with the help of aeration devices and recirculation of the water systems and proper feedings, it has become possible to stock 1.5 to 2 crores of spawn/ hectare with a high percentage of survival. In the case of fry, the general stock rate is 2.5-2.0 lakhs per hectare but under modern technologies, it is possible to increase the stocking density.

Counting of Spawn/Fry

Spawn are measured by measuring cups which are perforated, cup-like structure of different size (5, 10, 20, 50, ml), made of metal. Two or three samples of spawn are taken and the number of spawn in each sample is counted. Then the mean of the samples is derived and that is the number of spawn in the cup. In the case of fry, the same procedure is followed, but the size of the cup is bigger. The general stocking rate of spawn of carps is 60 to 70 lakh/ha, and 6-8 thousand/ha, in case of fingerlings.

Material Required

- ☆ Measuring cups for spawn/fry.
- ☆ Tub, buckets, mugs, for transfer of spawn and fry.
- ☆ Watch glasses.
- ☆ Muslin cloth.

Procedure

- ☆ Calculate the stocking density as per the nursery or rearing pond area. Take out the spawn from the hatchery or other breeding grounds in a hapa and keep in a tub filled with water.
- ☆ Arrange a number of buckets with water (5-6 litres) according to the availability of spawn.

- ☆ Measure the spawn from the measuring cups and keep in the buckets (2 lakhs in each).
- ☆ Take the buckets to the nursery site.
- ☆ Enter the nursery pond up to knee level and dip half of the bucket in the water and then wait for 5 to 10 minutes.
- ☆ Keep the bucket in the water in such a way that the spawn/fry can get in the pond.
- ☆ Wash the bucket properly in the nursery pond.

Observations

The student should observe and record the following:

- ☆ Mortality of spawn or fry,
- ☆ Movement of spawn/fry along the periphery of the nursery ponds.
- ☆ The pupil should maintain a diary.

Precautions

- ☆ Take 3 to 4 samples of spawn and count to find out the average number of spawn/fry per measuring cup.
- ☆ Handle the spawn or fry carefully to avoid mortality.
- ☆ Stock the nursery ponds with spawn preferably in cool water.
- ☆ Do not release the spawn and fry at a time in one pond but release them separately into the nurseries/rearing ponds.
- ☆ If an oil layer is present on the surface after the application of oil emulsion in the nursery pond, remove it by splashing or disturbing the water or with the help of cotton cloth.
- ☆ Do no release all the spawn/fry at one place in the nurseries/rearing pond.

Exercise No. 5

Preparation and Application of Supplementary Feeds in Ponds

Aims and Objectives

- ☆ To understand the importance of supplementary feeds in nursery pond management and to understand the nutritional value of fish feed in growth..
- ☆ To distinguish between different supplementary feeds and to select suitable feed for better growth and survival of spawn, fry and fingerlings.

Relevant Information

Supplementary feed is that feed which is added to fish ponds in addition to the natural available there.

Type of Supplementary Feeds

Animal matter

- ☆ Zooplankton
- ☆ Fish meal
- ☆ Bone meal
- ☆ Silk worm pupae

Plant Material

Vitamins

Minerals

Oil

Yeast

Growth Promoting Substances

Method of Feed Preparation

Plankton Culture

- ☆ Crude culture of desired plankton can be prepared with Mohua oilcake, poultry manure and cow dung in the ratio of 6: 1:3 at the rate of 1,000 ppm.
- ☆ A culture of diatoms, (*Nitzschi* and *Navicula*) can be undertaken in the laboratory field with urea-bone meal and potassium nitrate in the ratio of 5 : 1 : 15 : 3.
- ☆ *Moina micrua* is cultured by using fresh cow dung and ground nut cake at the rate of 250-350 pm and 50 ppm, respectively.
- ☆ *Chlorella vulgaris* is cultured using NPK in the ratio of 8 : 9 : 5.
- ☆ Feeds in combination of different ingredients.

Exercise No. 6

Qualitative and Quantitative Analysis of Zooplanktons and Phytoplanktons

Aims and Objectives

To know the importance of plankton as food for spawn, fry and fingerlings and to understand the relationships of plankton production to various physico-chemical parameters.

Relevant Information

☆ Plankton is the small organisms of the animal and plant kingdoms which keep floating in the water, due to wave and current action. The plankton of the plant kingdom are termed as: "phytoplanktons". These are algae belonging to class *Bacilariophyceae* (diatoms) *Chlorophyceae* (Green algae) and *Cyanophyceae* (blue green algae).

☆ These groups of algae are found in all the waters in varying proportions. In muddy, polluted water with a high organic content in the soil, the blue green algae predominates; these are undesirable algae. In a nursery the most important algae are diatoms and green algae.

☆ The zooplankton comprises the copepods. Cladocera rotifers and protozoa. Of these, rotifers being small in size, as compared to cladocera and copepoda serve as food for the fry, in the early stages.

☆ By fertilizing the nursery ponds with organic and inorganic manure, the growth of plankton is encouraged and a consistently high population is maintained.

- ☆ In the intensive culture and rearing of spawn, the natural food in the nursery soon gets exhausted. To compensate for this depletion, the desired species of rotifers are cultured separately and introduced into the nursery pond.
- ☆ The planktons are collected from the nursery pond by a hand net. A hand net is 20 cm. in diameter with a conical net of bolting silk, having a mesh size of 120-250. At the end of the net a glass tube is fixed. Fifty litres of water are collected from different places in the pond and filtered through the net and, thus, the planktons are collected in the tube.
- ☆ The optimum level of planktons can be judged by filtering 50 liter of pond water through a plankton net. If the plankton is 1.0 cc in volume, it is considered that the nursery has an optimum level of plankton.
- ☆ The planktons are transferred to another tube and fixed in 4 per cent formalin solution. The plankton settles down within 24 hours.
- ☆ After they are transferred to a graduated tube and centrifuged. The plankton settles at the bottom and its volume is noted.
- ☆ An average of 3-4 sub-sample readings are multiplied by total plankton volume which will give the plankton quantity.
- ☆ For numeral and quantitative analysis 0.2 cc of plankton is kept on a Sedzwick-Rafter cell and all the organisms are counted under a binocular microscope and results o\ are endorsed on a data sheet.

Precautions

- ☆ While collecting planktons in the bucket, it should be submerged in the water, and taken out gently, otherwise the planktons will get disturbed.
- ☆ Allow sufficient time for the plankton to settle down.

Materials Required

- ☆ Buckets
- ☆ Plankton net

- ☆ Specimen bottles for plankton preservation
- ☆ Neutral 4 per cent formaldehyde solution
- ☆ Lugol's solution 6 gm potassium iodide mixed with 4 gm iodine in 100 ml of water.
- ☆ 500 ml plastic bottle with a wide mouth for phytoplankton preservation
- ☆ Dropper
- ☆ Measuring cylinder
- ☆ Hand centrifuge machine
- ☆ Rafter cell
- ☆ Binocular microscope
- ☆ Light microscope having 200 magnification.

Procedure

- ☆ Keep bucket, specimen tube, formalin, plankton net and a plastic bottle with a wide mouth near a nursery pond.
- ☆ Take a bucket full of nursery pond water from different portions of the nursery pond and filter through plankton net. (Total quantity of water from different areas should be 50 lt.)
- ☆ Take the plankton collected in the tube at the end of the plankton net into a specimen tube.
- ☆ Add formaldehyde solution to preserve the plankton.
- ☆ After the plankton settles down in the tube, decent excess water and transfer it to a 10 cc graduated tube.
- ☆ Centrifuge in a hand centrifuge machine.
- ☆ Note the volume of the plankton.
- ☆ Shake the tube for uniform distribution of the plankton.
- ☆ Take 0.2 cc of the plankton in a pipette and transfer it into the Sedzwick–Rafter cell.
- ☆ Cover with a large cover slip.
- ☆ Examine the organisms under a microscope and note down their number on the data sheet.

- ☆ Identify and record the plankton groups in general and individuals in particular, with the help of illustrations.
- ☆ Estimate the plankton qualitatively and quantitatively.

Exercise No. 7

Examination and Analysis of Stomach Content of Fish

Field Examination of Stomach

In a field examination of stomach the following preliminary notations are used.

Degree of Fullness

Full, half full, nearly empty, empty

Degree of Digestion

Fresh, half digested, digested

Type of Food in the Stomach

Give the identifiable species or groups of species and estimate their relative abundance (*e.g.*, fish larvae 50 per cent; Euphausids 25 per cent; unidentified 25 per cent).

Stage of the Fish (mainly applicable for Clupeoids)

- ☆ 1° Lean; no traces of fat in the digestive tube
- ☆ 2° Not very fat; a string of fat no thicker than 1 mm along the digestive tube.
- ☆ 4° Very fat; the digestive tube is completely enveloped by fat.
- ☆ Exact volumetric measurements of stomach contents and detailed identification of food species found in the stomach are usually carried out in the laboratory.

Stomach Analyses

Processing of Stomach Content

☆ The processing and analysis of stomach content of marine and freshwater fish are essentially the same. In the following pages are given instructions which apply strictly to fresh-water species and with slight modifications, also to marine species.

Carnivorous or Omnivorous Species

☆ Remove from the formalin jars the number of packets which can be completed in a regular work period and soak in water for 24 hours. This will remove most of the unpleasantness of the formalin. Caution must be used not to leave samples in water longer than 48 hours or they will start deteriorating. A deformalizing solution also can be used very satisfactorily.

☆ Place individual digestive tract in a Petri dish or other suitable container and remove all foreign matter (fat, liver, pancreases, etc.)

☆ Remove stomach, cut open, flush out organisms with bulb syringe or wash bottle. Some particles may still adhere to stomach walls and these should be picked carefully with forceps. Use as little water possible in washing out contents.

☆ Examine contents under dissecting microscope; separate, identify and enumerate organisms present.

☆ Place the individual organisms or groups of organisms on a blotter for not more than one minute to remove excess moisture.

☆ After this brief drying, place the various individuals or group of individuals in a centrifuge tube containing 5 cc of water and record volume displacement for each.

☆ Compute the percentage of total volume of each organism or group present.

☆ If there is a large amount of plankton present, pour sample into centrifuge tube and allow settling for 15 minutes. Record the volume of plankton. Then dilute to 5 or 10 cc and gently mix concentrate by inverting tube several times. With wide-mouth pipette immediately remove sample before plankters settle, and fill a Sedgwick-Rafter cell as for plankton analysis. Count organisms and multiply by dilution and record total numbers.

- ☆ When the volume of any organism is less than 0.1 cc record as T for trace.
- ☆ Use symbol <0.01 cc for less than 0.1 cc in grand total.
- ☆ Fish found in stomachs should be identified if possible. Also measure to the nearest millimeter if whole.
- ☆ A record sheet should be made out for each stomach examined, even if empty.
- ☆ Under comments include such information as: kind and number of parasites found, abundance of winter or summer eggs of copepods, etc.
- ☆ Preserve and label any organism not previously found.
- ☆ Stomach contents are identifiable or not. Do not clutter up record form with question marks.

Laboratory Analysis for Herbivorous Species

- ☆ Remove packets as in 1 above.
- ☆ Same as 2 above.
- ☆ Obtain a 2 cc sample from the most anterior part of the intestine. In some cases the entire contents of the whole intestine will be needed and the sample will still not equal 2 cc.
- ☆ Place 2 cc or entire sample if less than 2 cc in a centrifuge tube. Some packing may be necessary to remove air bubbles. A gentle stirring with a teasing needle should be sufficient for the purpose.
- ☆ If a 2 cc sample is obtained, dilute to 10 cc level. Stir and shake to break up any large clumps of some forms. If less than 2 cc is obtained, dilute proportionately. For example, if a 1 cc sample is obtained, dilute to 5 cc; if 1.5 is had, dilute to 7.5 cc etc. This will give a constant dilution of the concentrate.

☆ With a 1 ml wide-mouth pipette remove a 1 ml sample and place in a Sedgwick-Rafter cell as for plankton analysis, count organisms and multiply by dilution and record total number on record form.

☆ Occasionally large clumps of algae and other vegetable matter will be ingested Estimate the percentage by volume this material makes up of the total and remove from sample and follow above procedure. This food material which cannot be counted should be considered separately and given only a percentage basis. Other countable items are expressed in percentage by numbers. Thus, the sum of the percentage items counted will total 100 per cent. For example, a carp sample was made up of 50 per cent algal mass and 50 per cent countable forms. Of the countable forms, cladocerans made up 90 per cent and copepods 10 per cent.

Specification of Gut Contents

☆ The following outline of methods is based mainly on the review by Hynes, H.B.N. (1950). The term food item is not intended to refer to a species; it may be a taxonomically heterogenous group classes together by some other common attributes. It should be borne in mind by interpretation of the results that the method of each has an influence on the stomach content.

Occurrence Method

☆ The number of fish in which each food item occurs is given as percentage of the total number of fish examined.

☆ The total number of occurrence of all items is summed, scaled down to given percentage composition of diet. The importance of small numbers and small food items is magnified with this method.

Number Method

☆ The total numbers of individuals of each food item are listed.

☆ Expressed as a percentage of total number of organisms in all fish (not applicable to vegetable food, etc.). Small food items magnified in importance.

Volume and Weight Methods

- ☆ The volume or weight of each food item of total food in each fish is given as percentage of weight of fish. The volume of each item may be estimated or measured.
- ☆ Counts of food organisms are multiplied by the known average weights of individuals of each item.
- ☆ Mean index of feeding = $v.10^5/I^3$ where v= displacement volume of stomach contents and 1 = body length (lizuka *et al.*, 1954). This is an accurate method, but occasionally large items cause an incorrect picture of their real importance.

Fullness Method

- ☆ Estimate the degree of fullness of each gut by a ranking or points system. This is really a form of 4. The accuracy of this method depends on individual experience.

Points Method

- ☆ Each item in each gut is graded as "common", "frequent", etc., with allowance for the size of the organism as well as its abundance (one large is treated as equivalent to many small). Give rank numbers or points for each category, sum and scale to percentage of total. This is essentially volumetric. It is rapid, easy, and requires no special apparatus; it may be modified by taking fullness also into account, and does not give a spurious impression of accuracy.

Exercise No. 8

Survey and Study of Various Types of Fish Ponds and the Implements Used in Fish Farming

Fish culture in tanks, though on a limited scale, has been practiced in India since as early as 350 B.C. The industry apparently became more popular in later period. Tanks were regularly stocked with fish, and their fishing rights were leased out periodically. Fish culture is very popular in the states of Bihar, Bengal and Orissa. In other states too, there has been a marked awakening and interest in inland fish culture. At present, about 7, 00,000 acres of inland waters are under fish culture in our country. This area; is very small as compared to the actually cultivable inland waters. In Chennai (Tamil Nadu) alone, there are over 8,50,000 acres of inlands waters, including estuaries and backwaters. There are over 40,000 tanks of widely varying areas. At present, in Haryana State, about 3,000 acres of inlands water area are under fish culture practices. In India, as whole, irrigation canals exceed 1, 00,000 km in length and wells are about 13.5 million in number. Ponds and tanks are more numerous in Bengal, Bihar and Orissa than in Madras or in any other States.

Cultivable waters are of two types, perennial or seasonal, and fishes can be cultivated in ponds, tanks, natural lakes, artificial reservoirs, backwaters, irrigation canals, wells, paddy fields, jhils, bheels, barrow-pits, quarry pools, swamps, and even short-season accumulation of water in shallow areas.

Layout Plan for a Typical Fish Farm

Dependable sources of water supply to ponds and soil, which will retain water, are of fundamental importance in the selection of a site for pond construction. Low lying, water logged or marshy areas are much suited for their excavation into fish ponds, but the soil, if peaty, would reduce the productivity of waters. A clay loamy soil is considered as the best from the point of view of water retention and fertility. When it is intended to rear carps in new ponds, the importance for the construction of Nursery, Rearing and Stocking fish ponds should be remembered. A small fish farm of 10 acres in extent should have the area, divided as follows for the construction of different types of fish ponds.

Type of Ponds	*Area (Acre)*	*No of Ponds*	*Size (ft.)*	*Level of Water (ft.)*
Nursery	0.5	20	40 × 20 × 5	3.5–4.5
Rearing	2.0	8	80 × 40 × 6	4.5–5.5
Stocking	7.5	4	Any length × 100 × 8	6.0–7.5

The nursery ponds should preferably be small and shallow, with 3 to 4 feet depth of water. Rearing ponds should be slightly larger, but need not be proportionately deep. Stocking ponds should be deeper and longer (six to eight feet and 250 to 300 feet, so that the fish will have a longer run for a healthy growth. The width of the stocking pond should not exceed 100 feet, so that relatively smaller sized fishing nets and a limited number of fishermen would be sufficient to net the fish. Large ponds are not only difficult for fishing but also difficult to manage, particularly in the control of the overgrowing weeds.

Shallow waters are more productive than deeper waters. If there is a regular and well assured supply of water in the nursery ponds, it may be kept very shallow (2 to 2.5 feet), as in a rice field. The bund should, however, be properly tufted to prevent erosion. Fully protected inlets and outlets should be provided in fish ponds, wherever necessary, to regulate the inflow and outflow of water.

The side and bottom of the ponds should, gently be sloped towards the outlet of the pond, so as to facilitate the complete draining of water, whenever desired. If a deeper pit or depression is provided near the outlet in the pond, all the fish in pond would automatically be collected there, when the pond is completely drainded.

Various Types of Fish Farm Implements

The farmer owning a small 10 acre fish farm, which consists of small nursery ponds, rearing ponds and long rectangular stocking ponds, should keep the following implements ready for routine use (Figures 7–10).

Sl. No .	*Name of Item*	*Quantity Required*	*Purpose*
1.	Drag net, very fine nylon mesh, 36 feet size, with floats and weights on the opposite sides of net	4	Netting and collecting fish fry, a week or 10 days old from nursery ponds. For their use in larger ponds, the two may be joined together
2.	Drag net, 0.25 inch nylon mesh, 36 feet x12 feet size, with floats and weight on the opposite sides of net	4	Netting advanced fry or early fingerlings from nursery ponds and for sorting and stocking in other ponds
3.	Drag net, 0.5 inch nylon mesh, 50 feet × 14 feet size, with floats and weight on the opposite sides of net	6	For periodic netting of fingerling for transferring to stocking ponds, or for giving them an occasional run and exercise in ponds
4.	Drag net, 1.0-1.5 inches nylon mesh, 50 feet × 16–20 feet size, with floats and weights on the opposite sides of net	6	For catching grown up fish and examining their maturity condition and growth periodically or netting for marketing
5.	Hapa: A miniature cloth tank of fine mesh, made up of nylon. 6 × 3 × 4 feet, with strong tapes at all the four corners, top and bottom	6	For temporary storage of fish after netting from ponds, before sales, necessary also for sorting fingerlings of carps from other varieties of fishes
6.	Breeding hapa: A miniature cloth tank, fine mesh, rectangular shape, 6 × 3 × 4 feet outer size, with 4 × 2.5 × 3 feet, another rectangular cloth tank (breeding tank) of mosquito net, fitted inside the outer hapa	12	For then breeding of Indian Major Carps and Common Carp by induced breeding technique/simple techniques and also breeding the Chinese Carps.

contd...

Sl. No.	*Name of Item*	*Quantity Required*	*Purpose*
7.	Cast Net: 0.25 inch nylon mesh, 6-8 feet diameter, with 10-15 feet rope at one end and lead beads weights (sinker) at the periphery of the net	6	For taking random samples of fry/ fingerlings in the nursery/ rearing tanks for the routine checking for health and growth
8.	Cast Net: 0.5 inch nylon mesh, 8-10 feet diameter, with 10-15 feet rope at one end and lead beads weights (sinker) at the periphery of the net	6	For taking random samples of advanced fingerlings from rearing ponds, for the routine checking for health and growth of fingerlings
9.	Cast Net: 1-1.5 inch nylon mesh, 15-20 feet diameter, with 10-15 feet rope at one end and weights at the periphery	6	For taking random samples of adult fishes from stocking ponds, for the checking of fishes for marketing or to sort out brood fishes for breeding purposes
10.	Wide mouthed earthen pots or hundies, about 5 to 6 gallon capacity of water (tins preferable, but costly)	6	For temporary storage of live fish fry, fingerlings or adults, after nerving and for transport fish seed from one place to another
11.	Buckets	6	For carrying fish seeds, manure etc. from place to place at fish farm
12.	Baskets	12	For carrying sand, silt, manure at the fish farm
13.	Spades	4	For trimming embankments, desilting of fish ponds
14.	Jack knife	4	For trimming aquatic weeds on the embankments and in the marginal areas of ponds
15.	Bamboo screen 8 to 10 feet long	144	For fixing the hapas in fish ponds and for other miscellaneous use during breeding season of fishes
16.	Split bamboo screen: 10 × 4 × 4 size, rectangular tank/ox of bamboo strips	6	It is use as a miniature conditioning tank or live pen in ponds, for fish to be kept alive after catching, but before marketing

contd...

Sl. No.	*Name of Item*	*Quantity Required*	*Purpose*
17.	Deweeder or bottom raker: (a wooden or metal beam), 6 × 8 feet long, with strong bamboo or metal spikes fixed in all the directions and with strong rope, tied at either end for dragging	1	Deweeding deeper areas of ponds by repeated dragging in all the direction, and removing the weeds, caught among the spikes every time the weeder is hauled; and disturbing the bottom mud to remove poisonous gases accumulating therein like. Ammonia, Nitrogen and Carbondioxide.
18.	Straw rope, loose but thick and about 100 feet long	2 bundles	Dragging the surface water to remove floating weeds. (Pistia, Lemna, Azolla, Wolffia), thick algal scums, etc., which often occur in great abundance in ponds.
19.	Small boat or float	1	Deweeding purpose, particularly when the pond is deep and broad, and abounds in rooted weeds like water lilies, which have to be plucked out.

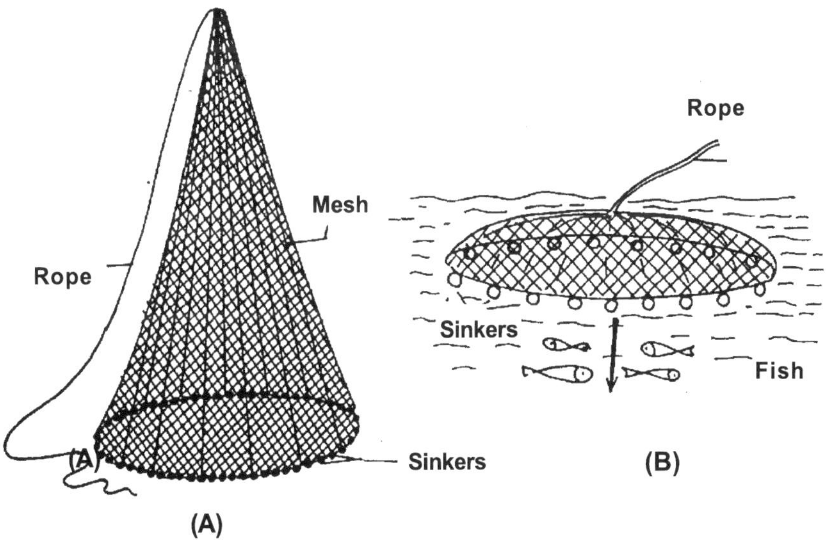

Figure 7: Cast Net
(A) Closed Cast Net; (B) Open Cast Net in Water

(A)

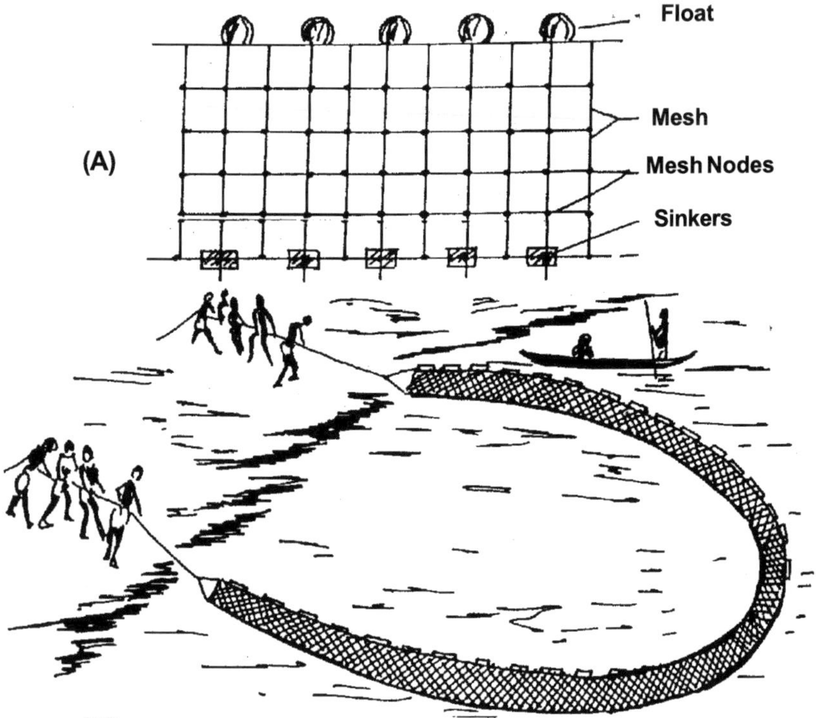

(B)

Figure 8: Drag Net
(A) Physical Construction of Drage Net
(B) Use of Drage Net a Water Body

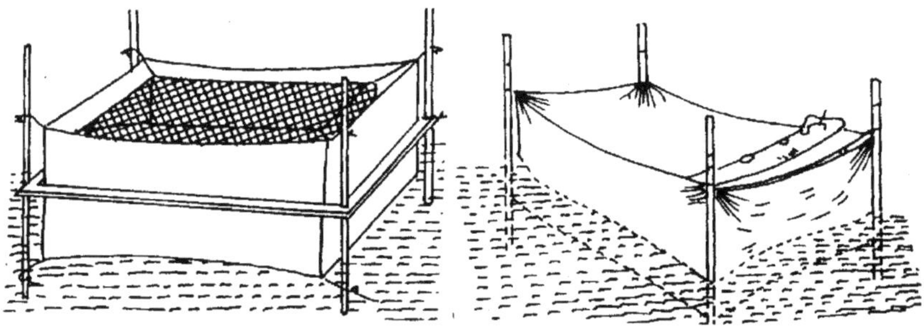

(A) (B)

Figure 9: Hapa Net
(A) Hatching Hapa; (B) Breeding Hapa

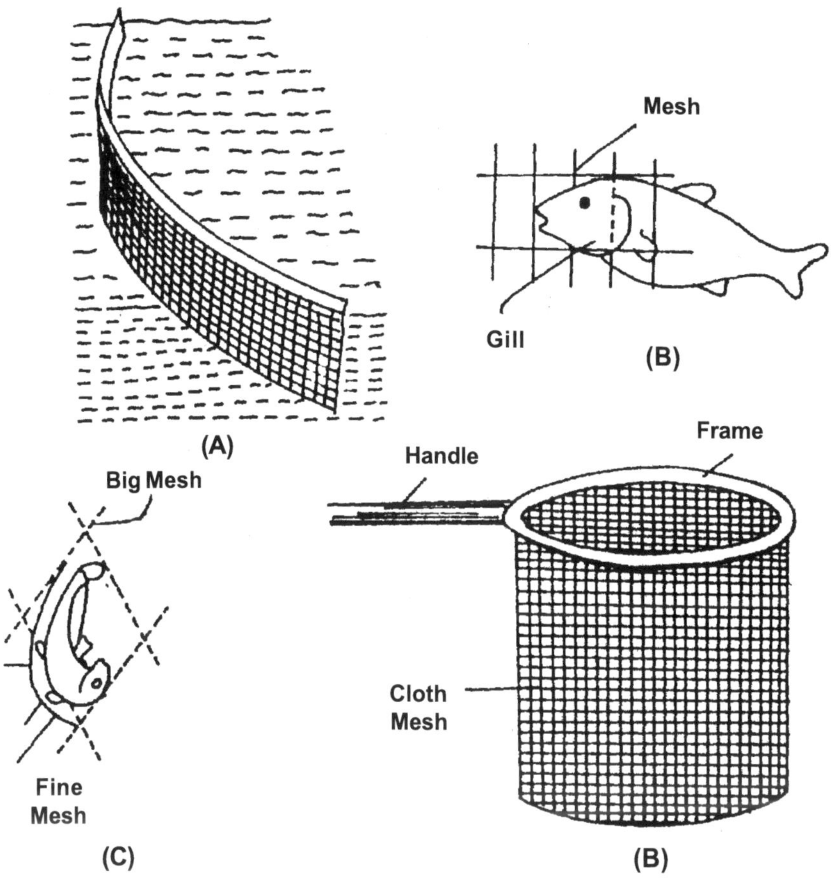

Figure 10: Other Types of Nets
(A) Sein Net (Patty Jal); (B) Fish Caught in a Sein Net; (C) Fish Caught in a Drag Net; (D) Hand Net

Some other miscellaneous items may also be required occasionally like an extra stock of wire netting for protecting the inlets and outlets of ponds, if any. A balance with weights for weighing fish, a scale (in inches and centimeters) for measuring their length, etc. will be necessary. For the proper maintenance, watch and ward of the fish farm will require the full-time service of at least two persons, who could also profitably be engaged in vegetables and floriculture farming on the banks and bund of fish ponds.

Exercise No. 9

Management of Fish Ponds

(A) Liming and pH Adjustment in Nursery Ponds

Aims and Objectives

- ☆ To understand the importance of liming in nursery ponds.
- ☆ To know the form in which lime is applied.

Relevant Information

- ☆ What are the important liming materials used in nursery pond management?
- ☆ Generally calcium oxide (CaO) called slaked lime or quick liming and calcium carbonate ($CaCO_3$) called chalk or lime stone are used as the liming material.
- ☆ Why liming is done in nursery pond?
- ☆ It is done to supply calcium which is an important nutrient. It not only acts as a buffer but also keeps the pH value on the alkaline side. By its application, acidic agents are neutralized. Moreover, it also acts as a disinfectant and kills harmful bacteria, and keeps the water clean. It can be used for the control of parasitic fish diseases also.
- ☆ It also releases certain nutrients like potash and phosphate from the bottom sludge into the water which go into the production cycle.

☆ The dosage of lime is calculated, based on the prevailing pH value of the pond.

pH	Pond Condition	Dosage of Lime (kg/ha)
4-4.5	Highly acidic	1000
4.5-5.5	Medium acidic	700
5.5-6.5	Slight acidic	500
6.5-7.5	Near neutral	200

☆ Pond water having a pH value 7.0 is neutral. When it is less than 7.0, it is acidic and when it is more than 7.0, it is alkaline. Acidic waters are unproductive. Hence, the pH of the pond water should be maintained on the alkaline side. If the pH value of the pond water is around 7.5 to 8, the pond is considered as highly productive. But if the pond water is near neutral, further liming is done as it increases the productivity of the pond.

Precautions

☆ Apply the liming material homogeneously.

☆ Use liming material in powder form.

☆ See that liming is done before fertilization.

☆ Use hand gloves while liming.

Procedure

☆ Find out the pH value of the pond with the help of litmus paper or pH strips.

☆ Calculate the quantity of lime required.

☆ Put on the rubber gloves.

☆ Weigh the liming material and convert it into powder.

☆ Take the powdered lime in a bucket and spread it evenly in the pond.

☆ Wash your hands thoroughly.

☆ Find out the pH value after liming.

(B) Manuring and Fertilization of Nursery Ponds

Aims and Objectives

- ☆ To understand the importance of manuring of nursery ponds.
- ☆ To distinguish between different manures and fertilizers.

Relevant Information

- ☆ Manuring is an important step in the management of nursery ponds for better growth in the rearing of carp spawn. Organic manures like cows dung and various oilcakes have been reported to increase production of phytoplankton and zooplankton which constitute the most suitable food items for carp hatchlings. But this involves problems in the procurement, transport and handling of large quantities of manures. Inorganic fertilizers, in smaller quantities, further help in enhancing the growth of phytoplankton and subsequent development of zooplankton. Therefore, a combination of organic manures and inorganic fertilizers is preferred in proper nursery management as it saves bulk handling.
- ☆ Manuring is the process of application of inorganic or organic materials providing nutrients to the water as well as to the soil. Manuring boosts the plant and animal life in the nursery pond as it ultimately supply food for the fish seed.

Types of Manures and Fertilizers Used in Nursery Ponds

- ☆ Organic manures
- ☆ Green manures
  - ☆ Animal dung
  - ☆ Groundnut oilcake/other edible oilcakes.
- ☆ Inorganic fertilizers/ chemicals
- ☆ Lime stone
- ☆ Phosphatic fertilizers (P_2O_5)
- ☆ Potassium fertilizers (K_2O)

- ☆ Nitrogenous fertilizers
- ☆ Cobalt chloride

Recommended Doses of Manures and Fertilizers

Organic Manures

Green Manure

Grass or mangrove leaves: These are applied @ 1775 kg/ha.

Animal Dung

Cow dung: It is applied @ 10,000 kg/ha, two -three days before the spawn is stocked and @ 5000 kg/ha, 7 days after spawn has been stocked.

Poultry droppings: These are applied @ 1,000 kg/ha, 10 days before spawn is stocked and 500 kg/ha, 7 days after the spawn has been stocked.

Pig dung or house dung: It is applied @ 800 kg/ha, 10 days before spawn is stocked and 400 kg/ ha, 7 days after spawn has been stocked.

Groundnut Oil Cake

It is applied @ 1000 kg/ha.

Inorganic Fertilizers/Chemicals

Lime Stone

Lime stone is a fertilizer material in which calcium as one of the essential plant nutrients. It is applied @ 250 kg/ha.

Phosphatic Fertilizers (P_2O_5)

The important phosphatic fertilizers which are commonly used in the nursery pond are single super phosphate and diammonium phosphate. The effect of phsophatic fertilizer application lasts for 2 to 3 years. These are applied @ 50 kg/ha.

Potassium Fertilizers (K_2O)

Potash is also one of the essential nutrients. It can be made available through Muriate of potash and potassium sulphate. These are generally applied along with other fertilizer materials @ 30 kg / ha.

Nitrogenous Fertilizers

Nitrogenous fertilizers often have a stimulating effect on the growth of phytoplankton which is one of the most important raw materials for fish production in ponds. The main source of nitrogen is urea or ammonia sulphate. It is applied @ 80 kg/ha.

Cobalt Chloride

It is applied @ 2kg/ha. Popular combination of manures and fertilizers used in nursery pond management: The combination are: (i) Single manuring: Raw cow dung 5,000 kg + groundnut oilcake 250 kg + superphospahte 25 kg + nitrogenous fertilizer 25 kg/ha.

Phased Manuring

- ☆ One day proceeding spawn stocking, superphosphatic lime 50 kg + cow dung or cattle dung 700 kg + oilcake 700 kg/ha.
- ☆ On the 2nd day of stocking: 350 kg oilcake + 87.5 kg cattle dung/ha.
- ☆ On the 3rd day of stocking: 175 kg oilcake + 43.75 kg cattle dung / ha.
- ☆ On the 4th to 9th day of stocking: Cattle dung 87.5 kg + oilcake 100 kg + superphospahte 22 kg /ha.

Oil Emulsion in the Nursery Pond

Due to manuring of ponds large number of aquatic insects develop in it. The main insects are swimmers (Notonecticdae), water boatman (Family: Corixidae), beetles larvae and bugs. Some of the insects also fly from pond to pond. As these insects prey upon the spawn, the survival of the spawn is reduced; it is, therefore, essential that the pond is cleared of the insect populations immediately before stocking of spawn and sometimes also about a week after stocking. The aquatic insects come to the water surface very often to breathe oxygen from the air through their tracheal tubes. The best method to kill these insects is by spreading an oil film on the surface of the water. Due to this operation, the tracheal tubes of the insects get chocked with the oil film and they are unable to breathe in oxygen from the atmosphere, and hence they die.

Methods of Application

Oil emulsion is generally applied by the following two ways:

- ☆ Mustard or coconut oil and cheap washing soap in the ratio of 56: 18 kg/ha.
- ☆ Light speed diesel oil 50 l/ha mixed with 1/3rd to 1/4th of its weight of cheap soap.
  - ☆ Oil or light speed diesel oil is put into a bucket, and a soap and water is added to it. Then it is thoroughly stirred till the entire solution becomes light brown in colour. It is then sprinkled on the water surface of the pond. After one hour, the insects die. The dead insects should be sieved out from the pond with a fine meshed muslin/nylon cloth of the size and depth of the pond so that all the dead insects are removed in one or two operations.

Table 1: Manuring of Nursery Ponds

Lime Stone (Kg)	*P_2O_2 (Kg)*	*Urea (Kg)*	*$Ca(No_3)_2$ (Kg)*	*K_2O (Kg)*	*Cow Dung (Kg)*	*Oil Cake (Kg)*
250	25	–	–	20	–	–
–	20	–	20	–	5,000	250
–	20	–	20	–	5,000	–
350	–	–	15	–	–	–
400	–	–	–	15	–	–
250	–	–	15	10	–	–
250	–	–	15	10	–	–
300	–	15	–	10		

Precautions

- ☆ Remove all the unwanted odours from the manures.
- ☆ Use the first dose of manure/fertilizer or mixture, 10-15 days before the spawn is stocked.
- ☆ Avoid over fertilization.
- ☆ Avoid fouling of water after fertilization.
- ☆ Check the growth of unwanted insect life which is harmful to the young spawn.

☆ Apply oil emulsion 12-24, hours before stocking of the pond in calm and still weather.

Procedure

☆ Measure the total area of water.

☆ Procure the previously done report on water analysis and the quantum of animal and plant life. In case it is not available, take the sample (soil, water and plankton) for analysis.

☆ Mix the required quantity of manures or fertilizers.

☆ Make a paste of the fertilizer mixture by adding the required quantity of water.

☆ Remove unwanted gases and odours from the fertilizer mixture by aeration.

☆ Broadcast the slurry/paste homogeneously.

☆ Apply oil emulsion 12-24 hours before stocking spawn in the pond.

☆ Apply the remaining quantity of manures of fertilizer mixture in phases, as recommended.

☆ Take water samples periodically and analyse them.

Observation and Calculations

The students should observe and analyse the following:

☆ Plankton growth from day to day.

☆ Physico-chemical conditions of the pond.

☆ Oxygen level.

☆ Alkalinity.

☆ Primary production of the ponds in terms of carbon.

☆ Periodic soil analysis.

☆ Growth and percentage of survival of spawn.

☆ Colour of the pond.

☆ Combine manures and fertilizers for better utilization and optimum growth.

☆ Work-out the phased manuring

Exercise No. 10

Identification of Common Aquatic Weeds of Freshwater Fish Ponds

Aquatic weeds are defined as those unwanted and undesirable plants which reproduce and grow in water and if left unchecked, choke the water body thus posing a serious threat to pisciculture, as they consume large quantities of soluble nutrients and drastically curtail the fish production. Weed choking of a body of water severely restricts plankton production; limits the living space of fish; upsets the equilibrium of physico-chemical qualities of water, chokes O_2 circulation, promotes accumulation of deposition of siltation; provides shelter to predatory and weed fishes, molluscs and aquatic insects and obstructs the netting operations.

Thus, weeds on the basis of their habits and habitats, having been variously classified. Lawrence (1955) grouped them into following types:

Floating Weeds (Figure 11)

Which have their free floating leaves above the surface of water with roots hanging free underneath, *e.g. Eichornia, Pistia, Spirodela, Lemna, Azolla* and *Wolffia.*

Emergent Weeds (Figure 12)

They are rooted in the bottom but having their floating leaves flush with the water surface and the stems are always hard, remain inside water body, *e.g. Nymphae (lotus), Nelumbo* and *Nymphoides.*

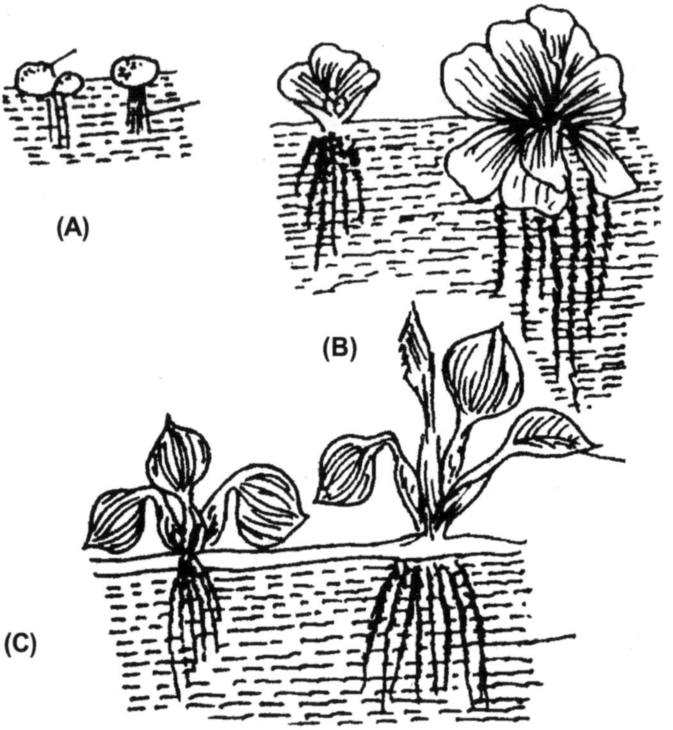

Figure 11: Floating Weeds
(A) *Lemna*; (B) *Pistia*; (C) Waterhyacinth

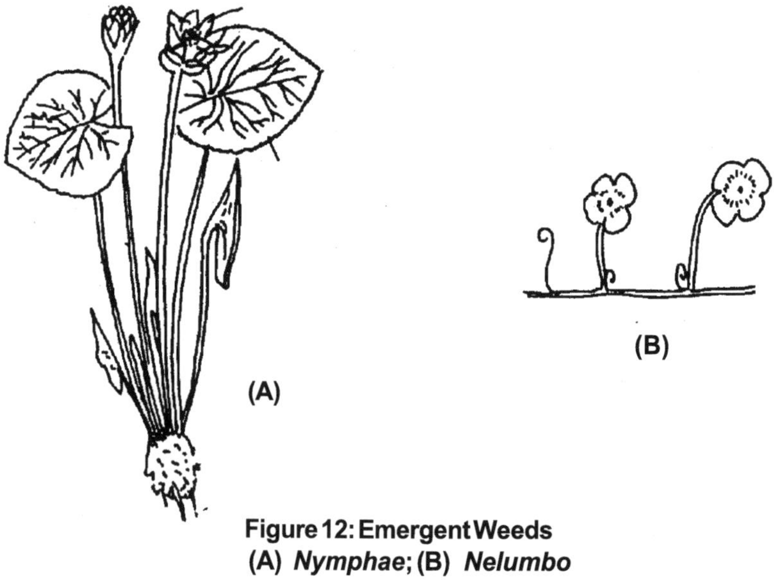

Figure 12: Emergent Weeds
(A) *Nymphae*; (B) *Nelumbo*

Submerged Weeds (Figure 13)

Which may or may not be rooted but whole of the body of the plant remains inside the water except the flowering part in some case, *e.g.* rooted *Hydrilla Najas, Vallisneria, Potamogeton, Chara* devoid of roots: *Ceretophylla, Utricularia, Najas*, etc.

Marginal Weeds

These are mostly rooted and choke the shore line of the water body, they are very hard and can susain without water for two-three months, *e.g. Typha, Pharagmities, Cyperus, Panicum, Colocasia, Ipomoea.*

Mats or Scum

They are formed by filamentous algae and some *Chlamydomonas* and Eugleninae family in the marginal area or in the main body of the water, thus form a thin film over the surface of water, *e.g. Spirogyra, Pithophora,* etc.

Algae Scattered through the Water Body

In this case these micro aquatic plants give greenish tinge to water body due to their presence when in abundance. They are scattered on the surface, column and bottom of the pond water, *e.g. Mycrocystis, Oscillatoria, Anabanena,* etc.

Phytoplanktons

These are minute microscopic plants passively floating in water, often multiplying very rapidly and occurring in millions imparting a turbid green, brown or yellowish brown colour to water, such a condition is known as *"water bloom"* and the plants are generally allied algae. The majority of planktonic algae in the Indian fresh-waters belong to the groups green algae (Chlorophyceae) and blue green algae (Cyanophyceae) though a number of diatoms (Bacillariophyceae) and flagellates (Euglenineae) are also commonly found.

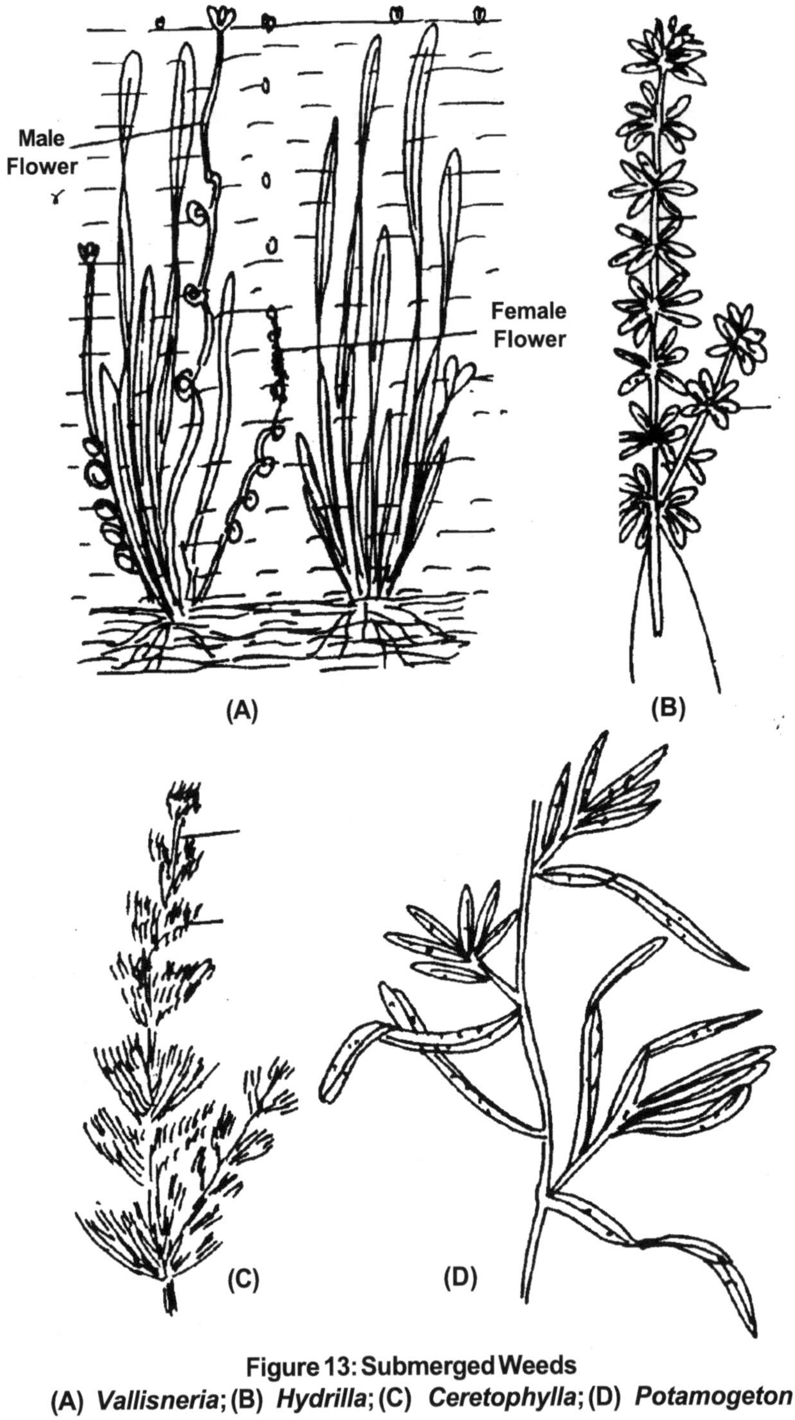

Figure 13: Submerged Weeds
(A) *Vallisneria*; (B) *Hydrilla*; (C) *Ceretophylla*; (D) *Potamogeton*

Exercise No. 11

Control of the Common Aquatic Weeds of Freshwater

After the pond is properly constructed, filled with water and fertilized, it supports a balanced aquatic vegetation requisite for pond ecology. When a new pond is constructed, the following precautions should be taken to check weeds from establishing themselves:

- ☆ Keeping minimum depth of water at not less than 0.5 m
- ☆ Keeping of the pond edge intact to reduce the hazard of marginal and shallow water weed growth.
- ☆ Shaping and sodding of the pond edge above water level to reduce the area where marginal weeds could appear.
- ☆ Providing a diversion or ditch, if necessary to carry excessive dirt and muddy water around the pond. The common methods adapted for control of aquatic weeds are: manual, mechanical, chemical and biological.

Manual and Mechanical Control

Manual methods are often employed to remove weeds. In small water bodies, the traditional methods of hand picking, uprooting emergent and marginal weeds and cutting them with scythes are considered suitable. The mechanical methods are relatively slow, rather expensive. In Bengal, a weed cutting winche having its bow fitted with long V shaped sickles and having reciprocal action, have been used in cutting weeds like *Colocasia, Cyperus* and other tall weeds in shallow water, power winches or mechanical lifts have been employed in Assam to clear shallow water thick infested with water hyacinth. Recently, diesel operated winches have been used in Orissa to eradicate dense rooted submerged vegetation.

Chemical Control

A large variety of chemical weedicides are now used for the control of aquatic weeds. Their actions, either by direct contact or by translocation of the chemical from the treated part of the plant result in death of the plant. Following points should be remembered before using herbicides.

- ☆ These should be effective in killing weeds at reasonably low rate of application, not exceeding 10 ppm, for those applied in water and about 10 kg/ha for those spread on a surface area basis.
- ☆ These should be cheap and easily available.
- ☆ These should be nontoxic to fish's human beings and stock animals.
- ☆ These should not pollute the treated water.
- ☆ These should not have an corroding action on metal container and sprayers.

Control of Floating Weeds

Among the floating weeds the water hyacinth (*Eichornia grassipers)* is most noxious and for its successful control 2, 4 diclorophenoxy acetic acid (2, 4-D) is used. This has direct effect upon the leaves and roots, if applied at the rate of 4.5–6.7 kg/ha. Other chemical is Taficide-80 (2, 4-D sodium salt 80 per cent) used at the rate of 4–6 kg/ha. It shows its effect within one week without an adverse effect on fingerlings and adults of major carps.

Other floating weeds such as *Pistia stratiotes* and *Salvinia* have been controlled by H.S.D and Powerine @ 775–1100 liter/ha, MCPA (2, 4-Dester) at the rate of 8.96–13.44kg/ha at 0.2 per cent concentration which are best for killing *Pistia*.

Control of Marginal Weeds

Young *Cyperus* are totally killed with 1 per cent aqueous solution of both Taficide-80 and Tafapon (2.2-dichloropropionic acid sodium salt) at the rate of 28 kg/ha. *Elocharis* is controlled with 2.5 per cent aqueous solution of 2,4-D with detergent at 0.25 per cent concentration, and *Colocasia* with Taficide-80 at a concentration of 1.5 per cent with detergent 'surf'.

Control of Emergent Weeds

In India, encouraging results have been achieved in the control of *Nymphae* with Dicotox and *Nelumbo* with Taficide-80 at a concentration of 1.5 per cent application of 2,4-D (sodium salt 80 per cent) at 1.5 per cent concentration with detergent surf. Taicide-10 at 2.5 per cent concentration with detergent showed good result on *nymphoides*.

Control of Submerged Weeds

Copper sulphate ($CuSO_4$) with ammonium sulphate (NH_2SO_4) is applied in controlling the *Chara*. Complete killing of *Hydrilla* and *Najas* were eliminated by using anhydrous ammonia. The gas injected directly in the top layer of the weeds chocked water by means of suitable application at the rate of 1.2 ppm. Other weedicides are Simazine at about 3-5 ppm active ingredient; also sulphur dioxide at about 40 ppm was found effective in killing *Hydrilla verticillate*.

Control of Algal Blooms and Filamentous Algae

The blue algae particularly *Microcystis* is most noxious algae. For this, Simazine was experimented at the rate of 0.5-1.0 ppm active ingredient, which could effectively control *Microcystis*. This weedicide is not toxic to fish or zooplankton.

Biological Control

Certain types of weeds in a body of water can be controlled by means of selected varieties of herbivorous fishes. The more important herbivorous fishes are grasscarp *(Ctenopharyngodon idella)*, Common carp, Tilapia. Gourami, an ideal herbivorous fish is one which consumes variety of weeds. Tilapia is best for *Chara* and *Najas*. Grass carp was found consuming wide variety of the aquatic vegetation.

Other animals such as duck and geese can be employed for controlling aquatic vegetation. In addition, they manure the ponds with their excreta, the geese keep the grass on pond banks trimmed short.

Exercise No. 12

Identification of Aquatic Insects of Fish Ponds

Aims and Objectives

To understand some of the important characteristics of the class Insecta and to identify common aquatic insects found in nursery ponds.

Relevant Information

- ☆ Nursery ponds are generally abundant in aquatic insects, most of which are harmful to the fish, either as predators or competitors. Hence, for the maintenance of healthy nursery ponds, it is necessary to identify and eradicate these harmful insects.
- ☆ Insects belong to the phylum Arthropoda which are characterized by the presence of jointed appendages (legs, antenna, antennules, etc.)
- ☆ The major characteristics of class Insecta is the division of the body into head, thorax and abdomen, presence of 2 pairs of wings and 3 pairs of walking legs. Respiration in this class of animals is by simple "air tubes" called "Trachae" traversing the entire length and breadth of the body and opening out by minute openings called "spiracles" The most commonly found freshwater insects are *Nepa, Corixa, Notonecta, Belostoma, Cybister, Gerris, Ranatra* and Dragonfly nymphs.

Precautions

- ☆ Dilute formalin to 4 per cent for preservation.
- ☆ Do not keep hands in formalin for long. It is better to use hand gloves while examining preserved specimens.
- ☆ After examination of preserved specimens, wash hands thoroughly with detergent soap.

Materials Required

- ☆ *Chat jal* (closed meshed drag net)
- ☆ Specimen bottles, watch glasses, glass slides, trays
- ☆ Dissection box
- ☆ Stage microscope
- ☆ 4 per cent formalin
- ☆ Sticker labels.

Procedure

- ☆ Net out nursery pond with chat jaal and collect the insect sample.
- ☆ Preserve the sample in 4 per cent formalin and take it to the laboratory for detailed examination.
- ☆ Transfer the sample to the tray and separate the different insects.
- ☆ Transfer the different insects to different specimen bottles and preserve in 4 per cent formalin.
- ☆ Undertake detailed examination under the stage microscope to identify the different insect, using the identification key given below.
- ☆ Count the number of insects in each group and find out the total count.

Key to Identification of Aquatic Insects (Figure 13)

Nepa

- ☆ Body blackish and dull in colour
- ☆ Head down ward in praying position.
- ☆ First pair of legs held in a raptorial position.

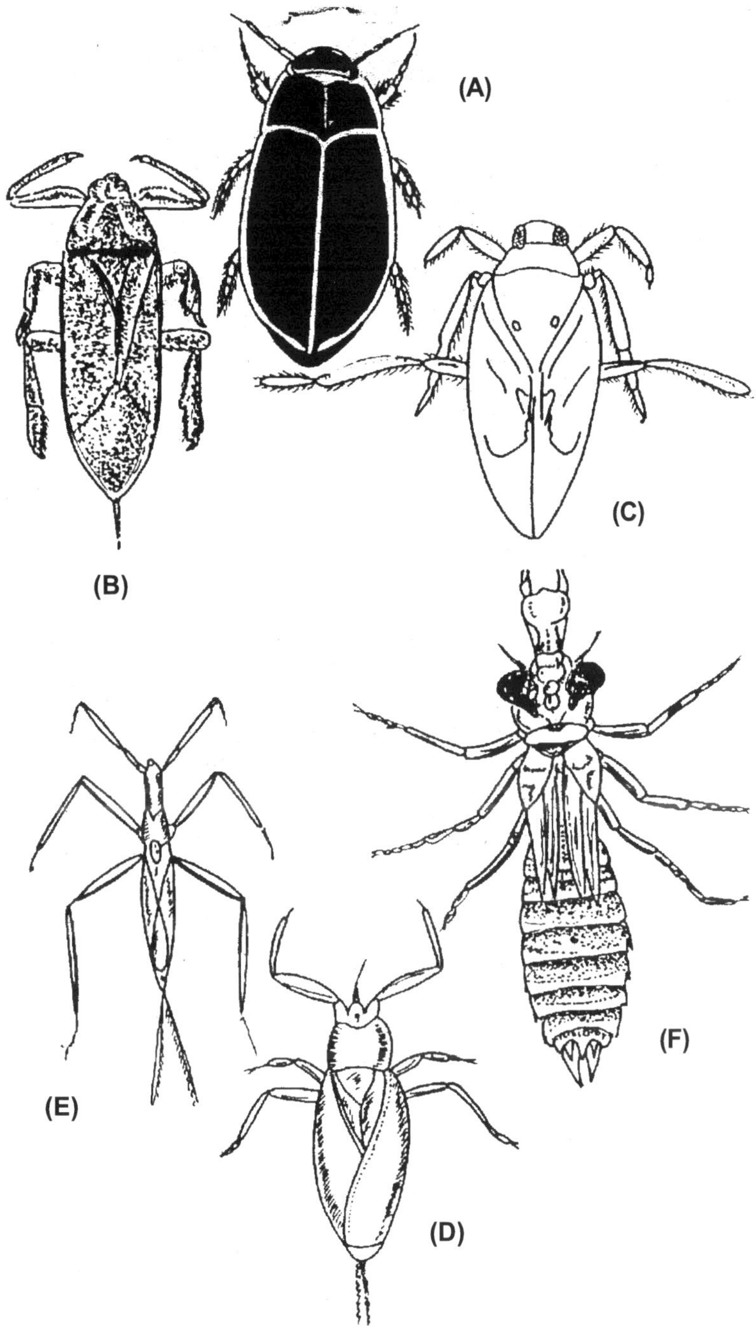

Figure 13: Predatory Aquatic Insects
(A) *Cybister* (Divine Beetle); (B) *Belostoma* (Giant Water Bug); (C) *Notonecta* (Back Swimmer); (D) *Nepa* (Water Scorpion); (E) *Ranantra* (Water Stick Insect); (F) Dragon Fly Nymph

- ☆ Two long respirator filaments or breathing tubes at the tips of the abdomen.
- ☆ First pair of wings covering the body.
- ☆ Thorax very flattened and prothorax narrow.
- ☆ *Nepa* commonly known as the water scorpion. This insect grows up to 17 mm in length. Because of their predatory nature, these insects are harmful to the fish fry.

Corixa

- ☆ Body dark greyish with faint yellow cross lines.
- ☆ Middle legs extraordinarily long.
- ☆ Hind legs flattened and fringed.
- ☆ *Corixa* commonly known as "water boatmen". They are very common in shallow ponds. They do not catch fry but they compete with them for food items like zooplankton and debris.

Notonecta

- ☆ Body bluish black in colour.
- ☆ Body deep, convex dorsally and keeled ventrally with sides sloping.
- ☆ Hind legs long and fringed for swimming, swim jerkily.
- ☆ Body has only 3 pairs of legs.
- ☆ *Notonecta* is commonly known as "back swimmers". These insects grow up to 10 mm in length and are harmful to spawn and fry in the carp nursery.

Belostoma

- ☆ Body brown or dull green in colour.
- ☆ Body large, flat and oval.
- ☆ Head bears a powerful beak
- ☆ Forelegs chelate and hind legs long and modified for swimming.
- ☆ Wing functional.
- ☆ *Belostoma* is commonly called "giant water bug". It grows to between 20-70 mm in length. These insects are predators in habit, attacking fish fry even larger than themselves.

Cybister

- ☆ Shining black or brownish black body, dull hollow margins.
- ☆ Head broad tightly fitting and capable of only slight movement.
- ☆ Antennae small and slender
- ☆ Hind legs long with hairs
- ☆ Wings functional
- ☆ *Cybister* is commonly called "diving beetle". It grows to 35 mm and has the biting type of mouth parts.

Cybister Larva

- ☆ Body muddy green in colour.
- ☆ Body long slender and segmented.
- ☆ Swims by up and down movements of the body. It can be seen hanging from the water surface film.
- ☆ *Cybister* larva is commonly known as "water tigers" or "dragons of the ponds".

Gerris

- ☆ Gregarious (live in group)
- ☆ Head small and body slender
- ☆ Forelegs small while last two pairs of legs 2-3 times longer than the body live on the water surface.
- ☆ The specimen is *Gerris* or "Water spider". They grow to 2-15 mm length

Ranatra

- ☆ Body dull green in colour.
- ☆ Prothorax narrower than head.
- ☆ Forelegs long and modified for grasping the prey.
- ☆ Two long filamentous respiratory tubes at the posterior end.
- ☆ Generally swims above the water surface.
- ☆ The specimen in *Ranatra* or "water stick insect". It grows

to 20-40 mm in length exclusive of the tracheal tube. Ranatra are very common insects in nursery ponds. The prey upon fish fry even up to 25 mm size.

Dragon Fly Nymph

- ☆ Body muddy brown in colour.
- ☆ Head broad, the lower side concealed by the development of the lower lip.
- ☆ Antenna short.
- ☆ Dragon fly nymphs are predatory in habit, consuming fish spawn and fries as well as other insects.

Unit–II

The scope of a animal physiology is wider than life itself. Animal Physiology is infact in the scene concerned with complete understanding of body functions at organ tissue or cell levels associated with living organisms over the part for decade, there has been phenomenon development in methodological technology both to understand and compare various life functions in the living systems. With the advancement in the knowledge of biochemical reactions and cell prepares, it is becoming increasingly difficult to encompass the current facts and theories about various life processes. Biochemistry and Physiology over top rather appeared merged rather irresparate to describe individually, the biochemist study the chemical constituents and the processes that occur in various organisms whereas a student of animal physiology will be profitably interested in major biochemical constituents and the rate functions.

In view of the above mentioned facts the brief knowledge, simple physiological tests would be essential to a fishery biology student to understand and compare the functional or the evolutionary status of various living organism or there chemical health. With this view, this unit is prepared to provide elementary knowledge of commonly applied physiological tests in animal, to the undergraduate or the postgraduate students.

Exercise No. 1

Methods of Unit Expression

Determining solution, concentration or strength is most fundamental aspect of the physiological studies in living system. Most commonly used system in chemical laboratories is as follows:

1. Percent by weight indicates the grams of solute per 100 ml of solution. 20 per cent of a solution means, 20 gm of solute dissolved per 100 ml of solution. Commercial HCl contain 38 per cent HCl by weight and has a density of 1.19. Each 100 ml of commercial product thus contain in 1.19 38 = 45 gm of HCl or the commercial HCl can also be expressed as 45 per cent by volume.
2. The mole or molar system of concentration gives the indicator weight of solute in a litter of solution. 180 gm of glucose per litre of a solution represent one molar concentration as per its molecular weight.

   Molar concentrations are normally indicated as 1M, 0.5M, 0.1M according to the number of gm moles prevent the litter of solution. Expression in brackets such as [1], [0.5]. [0.1] are often used to express molar concentration. Other measuring units are millimole (mM) *i.e.* 0.001M, micromoles (mM) *i.e.* 0.001 mM and miliosmol (m osmol). The term milli osmol is applied in connection with osmotic pressure in case of a non-electrolyte such as glucose 1 mmole is equal to 1m osmol, as osmotic pressure is proportionate to the total number of solid particles in a solution.

In case of electrolytes such as NaCl, 1mM of NaCl is equal to 2m osmol of Na^+ and Cl^-. Likewise one mM of Zn^{++} represent 2m equivalent and 1 mM of $A1^{+++}$ ion is equal to 3m equivalent as per their valancy.

Milli moles may be calculated by dividing milli grams per litre by the formula weight *i.e.* 78 milli grams of K^+ ion represent 78/39 *i.e.* equal to two millimoles or two milliosmols being mono valent. 100 milli gram of Cu^{++} ions per liter represent 2.5 milli moles (100/40), 2.5 milli osmol and 5 milli equivalent being a divalent, 222 milli gram of $CaCl_2$ per litter represent 2 millimoles (222/11) of calcium chloride and 6 milli osmol (2 Ca^{++} ion and 2 2 Cl^- ion).

3. Molar solution represents a gram molecular weight of solutes dissolved in 1000 gm of solvent. In a given solvent molar solution contain the same ratio of solute to solvent molecule irrespective to the molecular weight differences *i.e.* 46 gm of ethyl alcohol and 342 gm of cane sugar in 1000 gm of water, the ratio of alcohol molecule to water molecule will be the same as the ratio of sugar molecule to water molecule whereas in molar solution of these substances, the alcohol solution containing much more water relative to a molecule of solute then the cane sugar solution because 342 gm of cane sugar occupy, several times the volume of 46 gm of alcohol. Molar solution provides a definite ratio of solute to solvent molecule.
4. Equal volumes of molar solution *i.e.* HCl and NaOH exactly neutralize each other, whereas a given volume of 1M H_2SO_4 would neutralize the same volume of 2M NaOH because H_2SO_4 supplies 2 H^+ ion per formula weight. A normal solution of an acid contains the weight of a acid per litter of solution providing formula weight of H ions and of alkaline ions (OH^-) per weight.

Determining solution concentration or strength is most fundamental aspect of the physiological studies in a living system. Most commonly used systems in chemical laboratories are as follows:

1.1 Solution Concentration

Percent by weight indicates the grams of solute per 100 ml of solution. 20 per cent of a solution mean, 20 gm of solute dissolved per 100 ml of solution. Substances like proteins, nucleic acids etc. which do not have a uniformly defined composition, their concentration is expressed in terms of weight per unit volume rather than moles per unit volume. The unit of volume is liter, so all concentrations should be expressed with *l* (g/*l*, mg/liter, µg/*l* etc.). The term per cent is also quite commonly used. However, to avoid any ambiguity it is necessary to properly define the basis of per cent solution, as illustrated by the following example. A five per cent solution of acetic acid would mean:

5 g of acetic acid per 100 g of solution (w/w)

5 g of acetic acid per 100 ml of solution (w/v)

5 ml of acetic acid per 100 ml of solution (v/v)

Thus, 1 per cent (w/v) solution of casein would imply 1 g of casein dissolved in solvent to give a final volume of 100 ml of the solution.

1.2 Mass Fraction

The mass fraction of a component in solution is the mass of that particular component per unit mass of the solution. If W_A and W_B represent the masses of components A and B in a solution, then

$$\text{Mass Fraction of A } (W_A) = \frac{W_A}{W_A + W_B}$$

$$\text{Mass Fraction of B } (W_B) = \frac{W_B}{W_A + W_B}$$

1.3 Mass Percentage or % (w/w)

It is the weight of the component present in 100 parts by weight of the solution. In a solution containing 10 g sugar in 40 g of water, then

$$\text{Mass \% of Sugar} = \frac{10}{(10 + 40)} \times 100 = 20\%$$

1.4 Percentage by Volume or % (v/v)

It is the volume of the component in 100 parts by volume of the solution. In a solution containing 20 ml alcohol in 80 ml of water, the per cent volume of alcohol will be

$$= \frac{20}{(20 + 80)} \times 100 = 20\%$$

1.5 Parts Per Million (ppm)

This is generally employed for those solutions in which a susbtance is present in a very small quantity. It represents gram of a solute per million grams of solution or the gram of a solute per million ml of the solution.

$$\text{ppm} = \frac{\text{Mass of the Component}}{\text{Total Mass of the Solution}} \times 10^6$$

$$\textbf{or}\ \text{ppm} = \frac{\text{g or ml of Solute or Substance}}{\text{g or ml of Solution}} \times 10^6$$

Thus, 1 ppm of solution of NaCl in water represents

1 ppm = 1 mg NaCl/*l* of solution

or = 1 mg NaCl/1000 ml of solution

or = 1 ug NaCl/ml of solution

Commercial HCl contains 38 per cent HCl by weight and has a density of 1.19. Each 100 ml of commercial product thus contains 1.19 38 = 45 gm of HCl or the commercial HCl can also be expressed as 45 per cent by volume.

1.6 Molarity

This is the most common method for expressing the concentration of a solution in biochemical studies. The molarity of a solution is the number of moles of the solute dissolved per liter of the solution. A solution which contains 1 mole of the solute in one liter of the solution is called a molar solution. Molarity of a solution can be calculated as follows:

The mole or molar system of concentration gives the gram molecular weight of solute in a litter of solution. 180 gm of glucose per litre of a solution represent one molar concentration as per its molecular weight. Molar concentrations are normally indicated as 1M, 0.5M or 0.1M according to the number of gm moles present in the litter of solution. Expression in brackets such as [1], [0.5] or [0.1] are often used to express molar concentration.

$$\text{Molarity} = \frac{\text{Weight of Solute in g/l of Solution}}{\text{Mol. wt. of Solute}}$$

Say to prepare 0.1 M of NaOH (mol. Wt. 40), 40 0.1 = 4 g of NaOH is to be dissolved per litre of water.

1 M solution = 1 mole of the substance/L of solution.

= 1 mmole/ml of solution

= 1 µmole/ul of solution

1 mM solution = 1 mmole/*l* of solution

= 1 µmole/ml of solution

Other measuring units are micromole (µM) *i.e.* 0.001 mM and milliosmol (m osmol). The term milli osmol is applied in connection with osmotic pressure in case of a non-electrolyte, such as glucose 1 mmole is equal to 1m osmol, as osmotic pressure is proportionate to the total number of solid particles in a solution.

In case of electrolytes such as NaCl, 1mM of NaCl is equal to 2m osmol of Na^+ and Cl^-. Likewise one mM of Zn^{++} represent 2m equivalent and 1 mM of $A1^{+++}$ is equal to 3m equivalent as per their valancy.

Millimoles may be calculated by dividing milligrams per litre by the formula weight *i.e.* 78 milligrams of K^+ ion represent 78/39 equal to two millimoles or two milliosmols being monovalent. 100 milligram of Cu^{++} ions per liter represent. 2.5 millimoles (100/40), 2.5 milliosmol and 5 milli equivalent being a divalent, 222 milligram of $CaCl_2$ per litter represent 2 millimole (222/111) of calcium chloride and 6 milli osmol (2 Ca^{++} and 2 2 Cl^-).

1.7 Molality (m)

A solution which contains 1 mole of the solute disolved in 1 kg of the solvent is called a molal solution. Hence,

$$\text{Molality} = \frac{\text{Weight of Solute in g/l of Solvent}}{\text{Mol. wt. of Solute}}$$

It is important to remember that in a molal solution, the amount of solvent is 1000g. Thus in case of aqueous solution, 1 molal solution is obtained by dissolving 1 mole of the solute in 1000 ml (since Sp. Gravity = 1) of water. For example, for preparing 1 m Na_2CO_3 solution, dissolve 106 g of Na_2CO_3 (Mol. wt. of Na_2CO_3= 106) in one kg of water.

1.8 Normality (N)

The normality of a solution is the number of gram equivalents of the solute per liter of the solution.

A solution having one g equivalent of the solute per liter of solution is called 1N solution. Therefore,

$$\text{Normality} = \frac{\text{Amount of a Substance in g/L of Solution}}{\text{Eq. Weight of Substance}}$$

For preparing 0.1 N Na_2CO_3 (Eq. wt.of Na_2CO_3= 53) solution, dissolve 5.3 g Na_2CO_3 in a final volume of 1 *l* of solution.

It represents a gram molecular weight of solute dissolved in 1000 gm of solvent. In a given solvent molal solution contain the same ratio of solute to solvent molecule irrespective to the molecular weight differences *i.e.* 46 gm of ethyl alcohol and 342 gm of cane sugar in 1000 gm of water, the ratio of alcohol molecule to water molecule will be the same as the ratio of sugar molecule to water molecule, whereas in molar solution of these substances, the alcohol solution containing much more water relative to a molecule of solute then the cane sugar solution because 342 gm of cane sugar occupy several times the volume of 46 gm of alcohol. Molal solution thus provides a definite ratio of solute to solvent molecule.

Equal volumes of molar solution *i.e.* HCl and NaOH exactly neutralize each other, whereas a given volume of 1M H_2SO_4 would neutralize the same volume of 2M NaOH because H_2SO_4 supplies 2 H^+ ion per formula weight. A normal solution of an acid contain the weight of a acid per litter of solution providing formula weight of H ions and of alkaline ions (OH^-) per weight.

Examine

1. Calculate how many grammes of glucose are needed to make 100 ml of a molar solution ? (glucose mol. wt =180).

   A molar solution = mol. wt dissolved in 1 litre

   1 mol/l glucose =180 g/*l* =18 g/100 ml

2. Calculate how many millimoles or micromoles per millilitre are present in the following solutions: (a) 6 mol/ L urea; (b) 0.15 mol/*l* NaCl.

   a. Urea 6 mol/*l* = 6 mmol/ml

   b. NaCl 0.15 mol/*l* = 0.15 mmol/ml

   =150 µmol/ml

3. Calculate how many grammes of glycine are there present in 10 ml of a 20 mmol/*l* solution ? (Glycine mol. wt = 75).

   20 mmol/*l* glycine = 20 umol/ml

   10 ml contains 10 20 µmol=200 µmol.

   200 µmol contains 200 75 µg or 200/1000 75 mg = 15 µg.

1.9 Stock Solution

Stock solution of a substance is the one having a concentration many folds higher than that actually required in the experiment. Stock solutions are prepared of the substances that are to be used frequently and are stable at higher concentrations for several days and can be used after appropriate dilution just before use. It is sometimes convenient to weigh out a relatively large amount of the compound and prepare a stock solution of that compound from which small amounts can be withdrawn at convenience and added to solution of other components. The use of a stock solution thus cuts down on the amount of pipetting and at the same time reduces variability between a number of similar incubation mixtures, assay mixtures etc.

Frequently, a particular volume of a solution containing a desired concentration of a susbtance has to be prepared by using its stock solution. A simple mathematical equation can be used to calculate the volume of the stock solution needed to prepare a solution of required volume containing desired concentration of the compound:

$$N_1V_1 = N_2V_2$$

where,

N_1 = Concentration of the solution to be prepared
V_1 = Volume of the solution to be prepared
N_2 = Concentration of the stock solution
V_2 = Volume of the stock solution

For example, for preparing 150 ml of a solution containing 10 mM Tris from a stock of 75 mM Tris, the volume of the stock solution which should be used can be calculated as follows:

$$V_2 = \frac{N_1 \ V_1}{N_2} = \frac{10 \times 150}{75} = 20 \text{ ml}$$

Thus by taking 20 ml of 75 mM stock solution of Tris and making its final volume to 150 ml (by adding water or other solutions as per the requirement of an experiment) a solution containing 10 mM Tris is obtained.

Exercise No. 2

pH and Buffer Action

2.1 pH

The term pH is used to denote the concentration of hydrogen ions in cells and body fluid. The functional groups such as amino and carboxylic groups of the biomolecule dissociate at specific pH value and many of the biological and physical properties of these molecules depend upon this dissociation. The fundamental biological principal is that the consistency of the internal environment of the body in terms of ionic or molecular composition or temperature must be kept within relatively narrow limits, if animal health is to be maintained. It chiefly applies to maintenance of pH and the concentration of various electrolytes such as Na^+, K^+, Ca^{++}, Mg^{++} and PO_4^{++} and the total body water contents.

The term pH was introduced in 1909 by Sorensen who defined pH as the negative log of the hydrogen ion concentration.

$$pH= \log [H^+]$$

Low pH values correspond to high concentration of H^+ *i.e.* acids and high pH values to high of OH^-, as in alkalies. Acids are proton donar and bases are proton acceptor. The pH scale 0 to 14 covers the practical range of acidity and alkalinity of solutions commonly encountered. The reactions of the tissue/fluids depend upon their pH. For example the pH of normal blood serum or ECF in humans ranges between 7.35 and 7.45 *i.e.* slightly alkaline. If the pH of the blood goes down *i.e.* a state of acidosis exist and death ensures as a result of acidic coma. Conversely, when the pH is above 7.45, it causes alkosis and death ensure at 7.8 pH due to tetany. Decrease in pH of the body fluid may also cause hyperirritability with titanic contraction of muscles and ultimately death. A through knowledge of buffer system is thus indispensable in understanding the acid-base balance of the body.

The pH of the solution can generally be determined accurately by potential measurement with certain electrodes, mostly glass electrodes. The potential of an electrode changes with temperature hence require temperature compensation. The glass electrodes are particularly useful in all kind of biofluids and chemical reagents. Commercial glass electrode equipments are callibrated to read directly pH value. A variety of pH meters are now available for such measurements.

2.2 Buffer System

A buffer system is one that resists a change in pH on the addition of acid or alkali and constitutes conjugate acid-base pair. Most commonly, the buffer solution consists of a mixture of a weak acid and its conjugate base, *e.g.* a mixture of acetic acid and sodium acetate is a buffer solution. A well known physiological buffer system is the carbonate-bicarbonate system of blood.

The basis for functioning of a buffer with regards to its ability to resist change in pH can be illustrated by the following example. Suppose we have a buffer containing 5 ml of 0.1 M sodium acetate and 4 ml of 0.1 M acetic acid. Let us examine the effect on pH on addition of 1 ml of 0.1 N HCl to this buffer.

(*a*) Before Adding Acid

Total volume of buffer is (5 ml + 4 ml) = 9 ml

Conc. of acetate = 5/9 0.1 M = 0.5/9 M

Conc. of acetic acid = 4/9 0.1 M = 0.4/9 M

PK′ of acetic acid = 4.76

$$\text{Thus, pH} = 4.76 + \log \frac{0.5/9}{0.4/9}$$

$$= 4.76 + (0.097) = 4.86$$

(b) After Adding Acid

HCl provides H^+ which combine with the acetate ion to give acetic acid. Thus, it reduces the amount of acetate ions present and increases the amount of undissociated acetic acid, leading to an alteration in the salt/acid ratio and hence to a change in pH.

The final volume of buffer after acid addition = 10 ml

Conc. of acetate after acid addition $= 5/10 \quad 0.1\,M - 1/10 \quad 0.1\,M = 0.04\,M$

Conc. of acetic acid after acid addition $= 4/10 \quad 0.1\,M + 1/10 \quad 0.1\,M = 0.05M$

Thus, pH $= 4.76 + \log 0.04/0.05$

$= 4.76 + (-0.09) = 4.66$

Thus, the pH change marginally from 4.86 to 4.66 on addition of 1 ml of 0.1N HCl. However, if the same amount of the acid is added to pure water its pH will change from 7.0 to 2.0.

The most important is the distinction between strong acids and weak acids or strong bases and weak bases. Strong acids such as HCl and H_2SO_4 dissociate completely at low pH and strong bases like KOH and NaOH dissociate only at high pH. Buffers are of primary importance in regulating the pH of the fluids and tissues of living organism within limits, constant with life and normal body functions. They are widely used in the laboratory to control the pH of culture media for microorganism and tissues. Many chemical reactions including those catalysed by enzymes require pH control *i.e.* the solutions are appropriately buffered. Buffers of blood plasma, lymph, spinal fluid are HCO^-_3, H_2CO_3, HPO^-_4 and H_2PO_4, proteinase and proteins. The bicarbonate- carbonic acid buffer system is present in greatest concentration (about 0.025 M HCO_3 and 0.0012 M H_2CO_3) and is most important in maintaining blood pH around 7.4. Some of the acids being continuously produced in the body are sulphuric, phosphoric, lactic, aceto-acetic, beta hydroxy butyric etc. Buffers inactivate acids and bases thereby forming their salt, which are excreted.

In a dietetic patient 100 gm or more of beta hydroxy butyric and aceto-acetic acid per day may be produced. To prevent fatal acidosis, these are effectively neutralized by HCO_3-/H_2CO_3 system. Amino acid being amphoteric contain both acidic and basic groups and react with both alkalies and acids to form salts. A substance that can act both as an acid and as a base are termed as amptoteric. HCl in water is a strong acid due to formation of H_2O^+ and CH_3COOH in water is a weak acid, but it in ammonia acts as a strong acid due to the formation of NH_4 ions. The compounds which are easily ionized such as HCl freezes at low temperature.

2.3 Measurement of pH

By using pH indicators (natural or chemically synthesized organic compounds) an approximate value of pH of a solution can be obtained. The indicators dissociate like a weak acid on coming in contact with solution and give range of colours depending on pH. Dissociated and undissociated form of the indicator have different colours. pH papers coated with indicators of different pH range are commercially available. However, accurate pH can be measured using pH meter which measures e.m.f. of a concentration developed form reference electrode, test solution and a glass electrode sensitive to H+ ions. The combined electrode constituting a glass and reference electrode is dipped into the test solution for accurate measurement of pH.

2.4 Composition of Some Commonly Used Buffers

Buffers can be made as stock solution and these are diluted before use. Preparation of some of buffers frequently used in biochemical studies is given below. Unless otherwise stated, by following these procedures, 0.1 M buffer will be obtained.

(*a*) Acetate Buffer

Stock Solutions

1. 0.2 M acetic acid (11.55 ml/L)
2. 0.2 M sodium acetate (16.4 g of sodium acetate or 27.2 g of sodium acetate $3H_2O$ per L.

x ml of i + y ml of ii, diluted to a total volume of 100 ml

x	*Y*	*pH*	*x*	*y*	*pH*
46.3	3.7	3.6	25.5	24.5	4.6
44.0	6.0	3.8	14.8	35.2	5.0
41.0	9.0	4.0	10.5	39.5	5.2
36.8	13.2	4.2	8.8	41.2	5.4
30.5	19.5	4.4	4.8	45.2	5.6

(*b*) Phosphate Buffer

Stock Solutions

1. 0.2 M monobasic sodium phosphate (27.8 g in 1 L)
2. 0.2 M dibasic sodium phosphate (53.65 g of $Na_2HPO_4.7H_2O$ or 71.7 g $Na_2HPO_4.12H_2O$ in 1 L)

x ml of i + y ml of ii, diluted to a total volume of 200 ml

x	*Y*	*pH*	*x*	*y*	*pH*
93.5	6.5	5.7	56.5	43.5	6.7
90.0	10.0	5.9	39.0	61.0	7.0
85.0	15.0	6.1	16.0	84.0	7.5
77.5	22.5	6.3	5.3	94.7	8.0
68.5	31.5	6.5			

Exercise No. 3

Estimations of Micro- and Macro-molecules

Living cells contain micromolecules such as simple sugars, fatty acids, amino acids and nucleotides that serve as monomers for a variety of macromolecules as well as structural units of intracellular function. The sugars and fatty acids are chiefly energy molecules and certain monosaccharides also act as specific cell stimulator. The amino acids are often involved in the transmission of impulses and their deamination or decarboxylation give rise to many other compounds of biological importance. Excess or deficiency of these micromolecules may prove deterimental to cells. The composition of tissue or cells infact is a dynamic process *i.e.* there is continuous synthesis or degradation of various protoplasmic constituents. The composition of a given tissue at any time thus represents a balance between the synthesis and decomposition or intake and discharge of materials from the body. Animals exhibit differences in their body composition further as per their age, sex and physiological health.

The human body as a whole is composed of about 2/3 water and 1/3 solids. Later consisted of carbohydrates about 0.5 per cent, protein and lipids each constituting 15 per cent and minerals about 8.5 per cent, remaining 1 per cent being the other solids. Organic materials that serve primarily as food for the tissue within body are glycogen, glucose, fats, common amino acids and vitamins. The major structural organic compounds are proteins, phospholipids, glucolipids and cholesterol. The waste product of the tissue include substances like CO_2, urea, creatine and uric acid.

The state of nutrition and metabolic activity or the physical health of an animal markedly influence the composition of body tissue. In a poorly nourished animal, liver glycogen may be as low as 1 per cent and whereas the same animal fed on a high carbohydrate diet may have liver glycogen as high as 15 per cent. Inadequate dietary proteins and fats also result in spectacular variations in blood composition. Under pathological conditions, large amounts of fats appear deposited in the liver causing acute liver atropy. In progressive muscle distrophy, the muscles may show a marked decrease in creatine, ATP etc.

3.1 Carbohydrates

The carbohydrates commonly referred as 'Saccharides' mean sugars, occur in many form in animal tissue, tissue fluids and blood. Although a variety of sugars such as starches, sucrose, lactose or dextrin are taken into food and these eventually are converted to simple hexose sugars like glucose. Glucose is the primary sugar being utilized by the blood tissue chiefly as energy molecules. Human blood normally contain 60-100 mg of glucose in each 100 ml of it, except in diabetic conditions when it may reach upto 1000 mg per 100 ml of blood. Excess of glucose in blood is converted into glycogen (glycogenesis) in liver cells and if required again, the glycogen is broken down into glucose (glycogenolysis) to make up the loss of glucose. Fructose and galactose are other important hexose sugars.

Starch, glycogen and cellulose are the complex polysaccharide sugars available as food to the animals. Sugars in general behave as very weak acids and form salts at high alkalinity. All the sugars that contain free sugar group, in alkaline solution will behave as powerful reducing agent. As a consequence these readily reduce oxidising ions such as Ag, Hg, Bi, Cu and ferrocynide ($FeCN_6$) ions and the sugars are oxidised to complex mixture of acids. This reducing action of sugars in alkaline solution is utilized for both the quantitative and qualitative determination of sugars. The sucrose and trehalose are non-reducing sugars.

Glyceraldehyde is a triose and the most commonly known nucleic acid sugar is the ribose, a pentose sugar. Sucrose, maltose and lactose are the disaccharide sugars which occur respectively in sugarcane, germinating seeds and milk. Certain sugars occur combined with certain amino groups such as N-acetyl derivatives (*e.g.* chitin). Number of important biological substances like mucopolysaccharides, ascorbic acid, vitamin C, inositol, ABO blood group and keratin are other important sugar derivatives useful to animals. The mucopolysaccharides such as hyaluronic acid act as lubricant in synovial joints of bone, ligaments and tendon. Heparin is a blood coagulant in the blood plasma. The sugar estimations in blood plasma or various body organs thus constitute an important clinical index in physiological studies.

3.2 Proteins

The term protein is derived from the greek work 'Proteinos', meaning holding first place. All proteins are the polymers of about 22 types of amino acids. Proteins are essential substances as regulators (enzymes and hormones) of physiological functions and to serve as building block units for sub-cellular, cellular and many other organic structures. Proteins do not melt but decompose when heated and are soluble in different solvents. The proteins are classified as simple, conjugated or derivative proteins, depending upon their chemical nature and structure. The simplest proteins are albumin, globulins, scleroproteins and heptones. Albumin and globulins are important serum proteins. The conjugated proteins occur as proteins in combination with other carbohydrates (mucoproteins) or lipids (lipoproteins).

Major blood proteins are the oxygen carrying haemoglobin and a blood clotting protein, the fibrinogen. Collagen and elastin are present in connective tissues, where as rhodopsin and iodopsin are the visual pigments. Proteins in general constitute hairs, hails, bones, muscles, and ligaments in our body. Proteins may occur as enzymes, hormones, structural proteins, contractile protein, transport protein or the regulatory protein. Proteins are further classified as basic proteins or the acidic proteins due to their specific amino acids. The histones and protoamines are the basic proteins, where as phosphoprotein such as ceasin and vitelline are acidic in nature.

The proteins appear as cations in acidic medium and as anions in basic medium and as neutral at iso-electric pH. Proteins act as buffers at both side of the isoelectric pH. The body proteins are generally held at alkaline pH as negative ions due to isoelectric pH below the pH of blood. Proteins are easily precipitated either with heavy metal ions or any foreign protein molecule in an animal body. The extreme changes in pH and temperature destroy the functional activity of the proteins such as enzymes because of changes in their tertiary or quaternary structures, called denaturation. This is the reason that the most of the living organisms can not survive at temperature above 45°C.

3.3 Amino Acids

These are the monomer of protein. About 22 amino acids are known to occur in proteins. These are bounded together by peptide bonds formed in between the amino- and carboxylic groups of adjoining amino acids. These are classified both as per their ionic nature, that is neutral, acidic and basic amino acids or due to their structural configration as aromatic and heterocyclic amino-acids. All the amino acids, except glycine show optical activity due to their assymetric carbon atom. These also occur in dextro- or leavo- form.

The amino acids with mono-amino or mono-carboxylic groups are known as neutral amino acids, whereas those with diamino- and monocarboxylic groups are considered basic amino acids and those with one amino and two carboxylic groups are acidic in nature. The amino acids are generally water soluble and may vary in their taste. Some amino acids like the sodium salt of glutamic and *i.e.* sodium glutamate is used as a good flavouring agent in food. The free amino acids are useful to living organisms in a variety of ways, both as food, taste and for reproductive functions. There are specific essential amino acids for different organisms, the deficiency of which may cause either metabolic disorder or hamper reproductive or the breeding activities. Deamination of amino acids help in the synthesis of urea which is an important process in detoxifying ammonia in the living vertebrates. Histamine, also a derived amino acid is an important substance in allergic reactions which are caused by a variety of foreign sensitizing materials. Likewise creatine and a C-phosphate chemical act as a reservoir of high energy phosphate readily convertible to ATP molecules in muscles and other tissues.

Plant and some microorganism synthetise all the amino acids, whereas animals can not synthesis at least 7 amino acids, termed as essential amino acids. These are required to be present in diet and these include methionine, leucine, isoleucine, valine, tryptophane, threonine and phenylalanine. Deficiency of certain amino acids in certain animals like insects cause depriciation in metabolism growth and even their survival. Amino acids may also act as phagostimulant and a few like citrulline, ornithine and arginine participate in urea formation. The amino acid tyrosine is converted into thyroxine and adrenaline, tryptophan is utilized in the formation of a vitamin nicotinamide. Alkatonuria, phenylketonuria and albinism are some of the pathogenic alterations related to hereditary effects in tyrosine or phenylalanine metabolism. Tryphtophan is essential for maintaining appetite, plasma protein and Hb in animals.

3.4 Lipids

These are the greasy materials having large non polar hydrocarbon units. These are now-a-days extracted primarily with chloroform and methanol solvents and the extraction is made in an inert atmosphere. From an extract various categories of lipids may be separated chromatographically. The fats represent the bulk of stored energy producing food in the body of a living organism.

These are stored in adipose tissue which may contain as much as 90 per cent of fat as triglycerides. Phospholipids, lipoproteins, cholesterol and glucolipids are other important forms of body lipids. These however, are not stored in the adipose tissue, but are found in various other body organs as essential structural constituents. Liver regulate cholesterol level in blood. Excessive metabolism of fatty acids result in excessive secretion of ketone bodies in the urine (ketonuria). The keto bodies are chiefly aceto-acetic acid, B-hydroxy butyric acid and acetone derivatives from the non-enzymatic decarboxylation of aceto-acetic acid. The cholesterol is formed in many tissues and its chief root of excretion is via bile as bile acids. The synthesis of sex hormone also occur from the cholesterol.

The fatty acids are the monomers of lipids. These occur as saturated fatty acids in animal tissue and unsaturated (oils) in plants. The human fat is rich in oleic (39 to 47 per cent) and palmatic acid (21 to 25 per cent). Milk fat contain predominately palmatic (23 per cent) and octa-decenoic acid (29 to 34 per cent). Fats in animals chiefly serve as reserve food and as insulating and pedding material to support the organism. Waxes are the ester of certain fatty acids. Egg yolk and nervous tissue are rich in lipids.

Oleic acid, linoleic acid, linolenic acid and arachidonic acid are the unsaturated fatty acids, where as palmitic acid and stearic acids are common forms of saturated fatty acids. Saturated acids increase the cholesterol level in the blood and may lead to cardio vascular disorders, hypertension or arteriosclerosis.

Cholesterol is the most abundant steroid in the animal kingdom and is clinically very important to man. It is insoluble in water, acids or alkalies, but it is reality soluble in ether, benzene, chloroform etc. Cholesterol has an very important function when mixed with a fat or oil as it absorb large amount of water. Cholesterol is a poor conductor of electricity, hence act as a good insulator against electric discharge and acts as a good insolation material in the nervous tissue. The bile salts and sex hormone like testosterone, progestrone, estrogen and also cortisone and aldosterone are the important derivatives of cholesterol.

Free cholesterol is equally distributed between the cells and plasma while the esterified form occurs only in plasma. The normal serum cholesterol lies within the range 100-250 mg/100 ml. The average serum cholesterol is about 200 mg/100 ml at 25 in men and rises slowly with age, reaching a peak figure at an ages 40–50 and then declining. Women in general show a lower cholesterol level than men until the menopause, when the value rises above that found in men of the same age.

A high serum cholesterol of 300 mg/100 ml or more in young adults is a serious indication of coronary disease. High levels of as much as 25 per cent above normal are also found in nephritis, diabetes, myxoedema and xanthomatosis.

3.5 Nucleic Acids

Nucleic acids are important chemical constituents of genes in all living cells and occur in the nucleus and mitochondria. These possess the fundamental property of store, express and transmit genetic information, thus participating in reproduction and heredity. Changes in their structure can lead to mutations causing variations, may be of useful or harmful nature. Nucleic acids are catagorised into two principle classes according to the nature of the sugar, *i.e.* the deoxyribose containing Deoxyribonucleic acid (DNA) and the ribose containing Ribonucleic acid (RNA). DNA is associated primarily with the process of heredity and is a double helix structure. The RNA primarily involves in protein synthesis and is single stand structure. DNA is the polymer of nucleotides, which are consisted of a nitrogenous base *i.e.* adenine and guanine as purines and cytosine and thymine as pyrimidines, a sugar and a phosphoric acid molecule. A unit consisted of only a molecule of sugar and one base is termed as nucleoside. The nucleosides and nucleotides are further named as per their base molecule. In RNA uracil is present in place of thymine. The cells from a given animal species contain constant quantity of DNA per nucleus, therefore, determination of DNA in a tissue indicate the number of cells in that tissue. As per the Chargaff's rule, the total amount of purine is always equivalent to pyrimidines and the ratio of adenine and thymine to cytosine and guanine is species specific. A sequence of three bases constitute a triplet and forms the basis of genetic code which regulate the synthesis of proteins in living organisms.

3.6 Detection and Quantitative Estimations of Micro-Molecules

Preparation of Tissues

It is important to carry out the assays as soon as possible after the animal is killed. For example, process fish as follows:

Stun the fish, decapitate and collect the blood into a heparinized porcelain bowl. Rapidly remove the liver, gastrocnemius muscle or other organs and place on some crushed ice in a beaker. Dissect out the heart and drop this into beaker containing isotonic saline. Centrifuge the blood and remove of the plasma.

3.7 Detection of Sugars

Exp. 1: Benedict's Test for Reducing Sugars

Benedict's test is more convenient and this reagent is more stable. In this method sodium citrate functions as a chelating agent. Presence of reducing sugars such as glucose results in the formation of red ppt of cuprous oxide.

$$Na_2CO_3 + 2\,H_2O \qquad 2NaOH + H_2CO_3$$

$$2NaOH + CuSO_4 \qquad Cu(OH)_2 + Na_2SO_4$$

$$Cu(OH)_2 \qquad CuO + H_2O$$

$$\text{D-Glucose} + 2\,CuO = \text{D-gluconic acid} + Cu_2O \text{ (Red ppt)}$$

Reagents

1. Benedict's reagent: Dissolve 173 g of sodium citrate and 100 g of anhydrous Na_2CO_3 in 600 ml of hot H_2O. Dilute to 800 ml with water.
2. Dissolve 17.3 g of $CuSO_4.5H_2O$ in 100 ml hot water. Cool and dilute to 100 ml.
3. Add Reagent No.ii to Reagent No.iii slowly with constant stirring. Make the final volume to 1 L.

Methodology

Add 0.5-1 ml of the test solution or sample extract to 2 ml of Benedict's reagent (Reagent IV). Keep the test tubes in a vigorously boiling water bath. Observe for the formation of red precipitates whose appearance would suggest the presence of reducing sugars in the given solution or sample extract.

Exp. 2: Picric Acid Test for Reducing Sugars

It is another test for detection of reducing sugars. The reducing sugars react with picric acid to form a red coloured picramic acid.

Reagents

1. Saturated picric acid: Dissolve 13 g picric acid in distilled water, boil and cool.
2. 10 per cent Na_2CO_3

Methodology

Add 1 ml saturated picric acid to 1 ml of sample solution, followed by 0.5 ml 10 per cent Na_2CO_3. Heat the test tubes in a boiling water bath. Appearance of red colour would indicate the presence of reducing sugars in the sample solution.

Exp. 3: Moor Test for Glucose

Reagents

2 per cent NaOH

Methodology

Boil an aqueous solution of glucose in a test tube with an equal volume of 2 per cent sodium hydroxide solution. The solution will turn yellow and then reddish brown due to the formation of caramel, a condensation product of sugar.

3.8 Estimation of Sugars

Exp. 4: Estimation of Carbohydrates by the Anthrone Method

The anthrone reaction is the basis of a rapid and convenient method for the determination of hexoses, hexuronic acids, either free or present in polysaccharides. The blue-green solution shows an absorption maximum at 620 nm.

Reagents

1. Anthrone reagent (2 g/l in conc. H_2SO_4).
2. Glucose (0.1 g/l)
3. Glycogen (0.1 g/l)

Methodology

1. Add 4 ml of the anthrone reagent to 1 ml of a protein-free carbohydrate solution and rapidly mix. (Care: strong acid).
2. Place the tubes in a boiling water bath for 10 min, cool and read the extinction at 620 nm against a reagent blank.
3. Prepare standard curves for the glucose and glycogen solutions and compare them. The total body sugars are estimated following the method of Dubois *et al.* (1956).

Reagents

1. 5 per cent phenol
2. Concentrated H_2SO_4

Methodology

To 0.1 ml of the extract, 1.0 ml of 5 per cent phenol and 5.0 ml of concentrated sulphuric acid are added. The contents are shaken rapidly and allowed to stand for half an hour at room temperature. O.D. is read at 490 nm against a reagent blank. Standard curve is plotted using graded concentrations of D-glucose.

Exp. 5: Determination of Reducing Sugar by the Somogyi-Nelson Method

The sugar is heated with an alkaline solution of copper tartrate and cuprous oxide is produced, which reacts with arsenomolybdate to give molybdenum blue; the intense blue colour is then measured in the colorimeter. Protein must be removed before proceeding with the determination, and this is carried out using zinc hydroxide as the protein precipitant to give a neutral protein-free solution.

Reagents

1. Alkaline copper tartrate solution (Dissolve 4 g of $CuSO_4 \cdot 5H_2O$, 24 g of anhydrous Na_2CO_3, 16 g of Na–K tartrate (Rochelle salt), and 180 g of anhydrous Na_2SO_4 in water and make up to 1 litre).
2. Nelson's arsenomolybdate reagent (available commercially)
3. Stock sugar standards (glucose, fructose or maltose 1 g/l)
4. Working sugar standard (glucose, fructose or maltose stock solutions diluted 1 in 10 before use to give solutions containing 100 µg /ml).
5. Zinc sulphate 5 per cent (50 g/l solution of $ZnSO_4.7H_2O$).
6. Barium hydroxide 0.3N (0.3 mol/l).

Methodology

1. Pipette 0.1 ml of the test solution containing 50-150 µg of sugar into 1.5 ml of water and thoroughly mix.
2. Add 0.2 ml of barium hydroxide solution to the mixture, followed by 0.2 ml of aqueous zinc sulphate with thorough shaking and centrifuge.
3. Use 1 ml of the supernatant for further analysis.
4. Add 1 ml of the alkaline copper tartrate reagent to the sample and mix well. Place a marble on top of each tube and heat on a boiling water bath for 15 min.
5. Cool the tubes in cold water, add 1 ml of the arsenomolybdate reagent, and leave to stand for a minute until the effervescence ceases.
6. Dilute the blue colour with water to a final volume of 10 ml (or any other convenint volume) and read the extinction at 510 nm.

Exp. 6: Determination of Total Soluble Sugars by Ferricyanide Method (Calorimetric Method)

This method is based on the principle that sugars are oxidised by alkaline potassium ferricyanide and the amount of ferrocyanide produced is then measured either photometrically or volumetrically. The amount of the total sugars can also be determined colorimetrically by Anthrone method.

Reagents

1. Saturated neutral lead acetate solution
2. Sodium oxalate (solid)
3. Sodium carbonate (solid)
4. Conc. HCl
5. Potassium ferricyanide solution: Dissolve 0.5 g potassium ferricyanide in 1 L of distilled water. Store it in a brown bottle.
6. Carbonate-cyanide reagent: Dissolve 5.3 g sodium carbonate and 0.65 g potassium cyanide in 1 L of distilled water.
7. Ferric ammonium sulphate solution: Dissolve 1.5 g of ferric ammonium sulphate in 1 L of 0.05 N H_2SO_4.
8. Tungstate solution: Prepare 10 per cent (w/v) sodium tungstate in distilled water.
9. Standard glucose solution: Dissolve 100 mg glucose in distilled water and make up the volume to 100 ml.
10. N/12 H_2SO_4: Take 2.26 ml of conc. H_2SO_4 and make its volume to 1L with distilled water.

Methodology

1. For extracting the sugars, suspend 1 g of finely powdered oven dried sample in 40 ml of distilled water and heat in boiling water bath for 30 min. Centrifuge at 3000 rpm for 20 min. Collect the supernatant. Again suspend the pellet in 20 ml of water and repeat this extraction step 6-8 times till the supernatant is free of sugars. Combine all supernatants.

2. Add 1-2 ml of saturated neutral lead acetate to the collected supernatant and after properly mixing it for 15 min, filter it through Whatman No.1 filter paper and make the volume to 250 ml with distilled water. Excess of lead acetate is then precipitated out with solid sodium oxalate.
3. For determination of reducing sugars the above extract (step 2) is used directly. However, for estimating the amount of total sugars, take 25 ml of the above extract (obtained in step 2) and hydrolyse it with 5 ml conc. HCl for 8 min at 68°C. After cooling, neutralise the hydrolysed 10 *l* with solid sodium carbonate and make up the volume to 100 ml with distilled water.
4. To 1 ml of the preparation (step 2) for reducing sugars and that obtained in step 3 for total sugars, add 0.2 ml of N/12 H_2SO_4 and 9 ml of distilled water. Mix thoroughly, add 0.02 ml sodium tungstate and after shaking the contents allow to stand for 10 min. Centrifuge it at 3000 rpm for 20 min. Add 1 ml of ferricyanide solution to 1 ml of aliquot of this supernatant, followed by addition of 1 ml of cyanide-carbonate solution. Heat this mixture for 15 min in a boiling water bath and cool immediately under running tap water. Now add 5 ml of ferric ammonium sulphate. Mix the solution and after 10 min record its O.D. at 690 nm.
5. Prepare a reference curve of glucose over a range of 0-100 µg as standard, following the procedure described above (step 4).

Calculations

Vol. of extract taken after tungstate treatment

For colour development = a ml

O.D. of the extract = y

Assuming that from the standard curve, y O.D. units correspond to m µg of glucose.

Wt. of sample taken =1 g

Total vol. of extract = 250 ml

$$\text{Conc. of sugars in 1 ml extract} = \frac{m}{a} \times v \ \mu g$$

where,

v is the total volume of the supernatant after treating with tungstate

$$\text{Reducing sugar } \mu g/100 g \text{ sample} = \frac{m}{a} \times v \times 250 \times 100$$

$$\text{Reducing sugar \%} = \frac{m \times v \times 250 \times 100}{a \times 1000 \times 1000} = \frac{mv}{40a}$$

Similarly, calculate the per cent total sugar content of the sample after taking into account the dilution factors while processing the sample (as involved in step 3 of procedure) for estimation of total sugars.

Exp. 7: Enzymatic Method for Qualitative Estimation of Glucose in Blood Plasma Using Glucose Oxidase

Glucose oxidase is found in the growth of *Penicillium notatum*. It catalyses oxidation of B-D-glucopyranose to D-glucono-1,5-lactone, accompanied by formation of hydrogen peroxide. D-glucono-1,5-lactone is converted to D-gluconic acid. This enzyme is highly specific for B-D-glucopyranose. The reaction mixture for estimation of glucose contains peroxidase which utilizes H_2O_2 to oxidise o-toluidine to give a blue coloured product. These reactions are rapid but the colour is not very stable and the readings are recorded usually after 8 min. The sequence of events involved in this method is depicted below:

D-glucose + O_2 + H_2O Glucose oxidase H_2O_2 + D-gluconic acid

H_2O_2 + o-toluidine Peroxidase oxidised toluidine

(blue coloured product)

Reagents

1. 10 per cent (w/v) $ZnSO_4$ solution.
2. Isotonic sodium sulphate (93 mM): Dissolve 13.206 g of Na_2SO_4 and make the volume to 1L with water.
3. Sodium sulphate - Zinc sulphate reagent: Take 55 ml of 10 per cent $ZnSO_4$ solution (Reagent I) and add isotonic Na_2SO_4 (Reagent 5) to make the volume 1L.
4. 0.5 N NaOH

5. 0. 0.15M acetate buffer (pH 5.0)
6. Glucose oxidase: 30 International Units (one I.U.= I .U. mole of substrate transformed per min).
7. Peroxidase: 30 International Units.
8. O-toluidined: Dissolve 1 g of o-toluidine in 100 ml of absolute ethanol and store in dark bottle.
9. Combined o-toluidine reagent: Prepare a mixture containing the ingredients- 150 ml of acetate buffer (Reagent No. 8), 30 I.U. of glucose oxidase, 30 I.U. of peroxidase and 1 ml of o-toluidine solution. Store the mixture in brown bottle and it is stable for several weeks.
10. Glucose standard: Prepare a fresh solution containing 10 mg of D-glucose in 100 ml of water

Methodology

1. Pipette 0.1 ml of blood sample into 1.8 ml of sodium sulphate-zinc sulphate reagent in a centrifuge tube and add 0.1 ml of 0.5 N NaOH to it.
2. Centrifuge at 5000 rpm for 15 min. Take 1 ml of the supernatant and for blank use 1 ml of sodium sulphate-zinc sulphate reagent. Add 5 ml of the combined o-toluidine reagent and mix thoroughly.
3. Incubate the tubes at 30°C and exactly after 8 min read the colour intensity at 625 nm.
4. In another set of test tubes take suitable aliquots of glucose standard and proceed as described above (step 1-3) and prepare a standard curve.
5. From the standard curve determine the quantity of glucose in the sample tube and calculate the amount of glucose per ml of the blood sample.

Exp. 8: The Estimation of Tissue Glycogen

Glycogen is released from the tissue by heating with strong alkali and is precipitated on the addition of ethanol. Sodium sulphate is added as a coprecipitant to give a quantitative yield of glycogen. The polysaccharide is then hydrolysed in acid and the glucose released is estimated.

Reagents

1. Fish of the same age, sex and weight (four fed and four starved for three days).
2. Heparin anticoagulant solution from pharmacy in porcelain bowl.
3. Isotonic saline (NaCl, 0.154 mol/l)
4. Ketostix test papers.
5. Sodium cacodylate buffer (0.1 mol/l, pH 6.5) (Care: poison)
6. Saturated Na_2SO_4
7. Ethanol (95 per cent v/v)
8. HCl (1.2 mol/l)
9. Phenol red indicator solution
10. NaOH (0.5 mol/l)
11. Reagent for the estimation of glucose (Exp. 7)

Methodology

(a) Isolation of Glycogen

Accurately weigh the complete heart and muscle and about 1.5 g of liver. Place the tissues into a calibrated centrifuge tube containing 2 ml of KOH (300 g/l) and heat in a boiling water bath for 20 min with occasional shaking. Cool the tubes in ice, add 0.2 ml of saturated Na_2SO_4, and mix thoroughly. Precipitate the glycogen by adding 5 ml of ethanol (95 per cent v/v), stand on ice for 5 min, and remove the precipitate by centrifugation. Discard the supernatant and dissolve the precipitated glycogen in about 5 ml of water with gentle warming, then dilute with distilled water to the 10 ml calibration mark and mix thoroughly.

(b) Hydrolysis and Estimation of Glycogen

Pipette duplicate 1 ml samples of the glycogen solutions into test tubes calibrated at 10 ml, add 1 ml of HCl (1.2 mol/l), place a marble on top of each tube, and heat in a boiling water bath for 2 h. At the end of this period, add 1 drop of phenol red indicator and neutralize carefully with NaOH (0.5 mol/l) until the indicator changes from pink through orange to a yellow colour. Dilute to 5 ml with distilled water and determine the glucose content by the glucose oxidase method (Exp.7).

Calculate the grammes of glycogen per 100 g of tissue and remember that glucose exists as the glycoside form ($C_6H_{10}O_5$) in glycogen with mol. Wt 162, not 180.

Exp. 9: The Effect of Diet and Hormones on the Glycogen Content of Fish Liver

Glycogen is the reserve carbohydrate of animals and is present in the liver, skeletal muscle, kidney, and most other tissues, but is virtually absent from the brain. In a well fed animal, glucose is converted into glycogen in the liver (glycogenesis), but during starvation when glucose is no longer absorbed from the gut the liver glycogen is broken down to glucose (glycogenesis) and becomes depleted in about 24 h. Glucose synthesis and degradation is controlled by several hormones. The hormone adrenalin from the adrenal medulla is secreted in increased amounts during 'stress' situations and causes physiological changes in the animal to allow it to cope with emergencies. This effect together with changes in the sympathetic nervous system has been termed as the 'fright, flight or fight' mechanisms, since the animal is prepare to deal with the situation in one or more of these ways. Likewise, steroids such as cortisol and cortisone from adrenal cortex produce changes which are opposite to those of insulin. Glucose utilization by the tissues is decreased, carbohydrate breakdown is also diminished by the mobilization of depot fat and the increase in glycogen synthesis by gluconeogenesis from amino acids.

Reagents

1. Adrenalin (sterile solution for injection)
2. Tricholoracetic acid (50 g/l)
3. Ethanol (95 per cent v/v)
4. Ether
5. Iodine solution (5 mmol/l in 30 g/l KI).
6. Reagents for the estimation of glycogen (Exp. 8).

Methodology

(a) Preparation of Fish

1. Maintain the fish on a standard diet for one week, then divide them into three equal groups (A,B,C).

2. Feed groups A and B ad lib, but starve groups C and D for 48 h before the experiment.
3. Inject the fish in group B subcutaneously with adrenalin (10 μg/100 g weight of fish) and kil them 10 min later.
4. Eight hours before the fish in group D are killed, give the subcutaneous injection of cortisol (1.2 mg/100 g weight of rat)
5. Group A are fed controls and group C the starved controls.

(b) Isolation of Liver Glycogen

1. Kill the fish, dissect out the livers as quickly as possible, and record the weights.
2. Cut the liver into small pieces with scissors and transfer most of it to the homogenizer, but leave a small portion of known weight at room temperature to be dealt with later.
3. Add ice-cold trichloracetic acid (50 g/l) to the liver (approx. 10 ml/g of liver) and homogenize thoroughly.
4. Decant the homogenate, filter through a Whatman No. 54 filter paper, and collect the filtrate in an ice-cold flask.
5. Add half the original volume of trichloracetic acid to the sediment, briefly homogenize, and filter as above.
6. Remove a drop from each filtrate and add it to a drop of iodine on a porcelain tile.
7. Combine the filtrates and add 2 vol. of 95 per cent ethanol.
8. Stir the mixture and leave until the precipitate of glycogen flocculates. If any difficulty is experienced at this stage in precipitating the glycogen then add a pinch of sodium chloride and gently warm.
9. After centrifugation, remove the supernatant, disolve the precipitate in the minimum volume of water, and reprecipitate the glycogen by the addition of 2 vol. of 95 per cent ethanol.
10. Centrifuge the glycogen and wash the precipitate twice with 95 per cent ethanol, then a little ether.
11. Collect again by centrifuging, spread the residue on a watch glass, and dry in a dessicator.

12. Record the yield of material and store it until required.
13. Extract the small portion of liver which has been left at room temperature with trichloracetic acid and test the filtrate with iodine as above.
14. Finally, use the anthrone method (Exp. 4) to determine the amount of glycogen in your preparation and record the yield per gramme of wet weight of liver.

3.9 Detection and Estimation of Proteins

Proteins give colour reaction due to their peptide structure and presence of different amino groups in their molecules. Such colour reactions are important in the qualitative detection and quantitative estimation of proteins and their constituents *i.e.* amino acids. The ninhydrin test is most important in this context, ninhydrin react with free amino acids or proteins to give a blue colour.

Several methods of protein estimations are available. The most important methods are the nitrogen content estimations by the Kjeldahl's procedure and the calorimetric methods in which colour developing reactions are taken into account.

(*a*) Kjeldahl's Procedure

In this procedure, a sample of material is digested with boiling concentrated H_2SO_4 to which a catalyst such as copper or mercury salt is added. Sodium or potassium sulphate is also added to raise the boiling point of the digestion mixture. The digestion process causes oxidation and destruction of proteins and conversion of nitrogen to ammonia. Ammonia remains in acid-mixture as ammonium sulphate. The amount of ammonia nitrogen is determined by making the digest alkaline, followed by distillation of the liberated ammonia into standard acid. Titration of excess acid with standard base. Since most proteins contain 16 per cent nitrogen, multiplying protein nitrogen amount with 6.25, equals the weight of protein.

(b) Calorimetric Method

The ninhydrin and biuret reaction are the most important colour tests for proteins. The biuret test apparently is due to coordination of cupric ions with the unshared electron pair of peptide nitrogen and oxygen of water to form a coloured coordination complex. The complex contain two $CONH_2$ groups joined directly or indirectly. If urea is heated at 180°C, it release ammonia and form a biuret compound as follows:

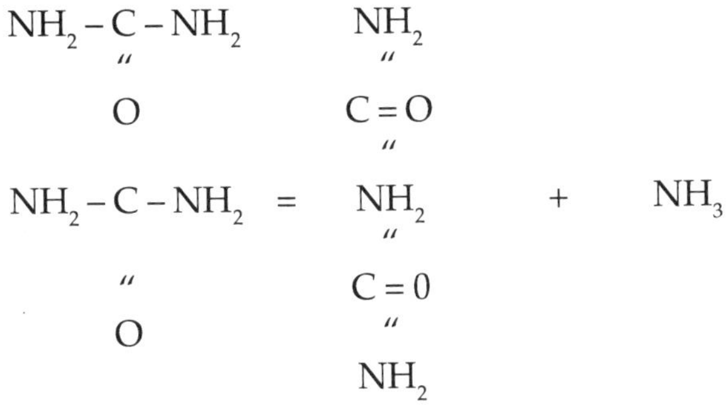

The presence of magensium sulphate or ammonium salt in the reaction mixture interfere with the biuret reaction. Addition of excess alkali minimizes their interference.

The biuret test is extensively used as a delicate test for the presence of proteins in biological materials. It has been converted into an excellent method for the quantitative determination of proteins in blood serum and other biological fluids.

(*c*) Xanthoproteic Test

Addition of concentrated nitric acid to protein solution generally causes the formation of a white precipitate which turn yellow upon heating. The colour appears orange when the solution is made alkaline.

(*d*) Biuret Test

In biuret test, the intensity of absorption is measured in the region 540 to 560 mμ for protein solution, in the presence of alkaline cupric tartrate and compared with the standard curve.

(*e*) Folin-cio-calteau Method

The final colouring in this method depends upon a combination of biurret reaction and reduction of phosphomolybidic phosphotungustic reagents by tyrosine and tryphophan present in the protein. This is most rapid and sensitive method.

Exp. 10: Detection of Proteins by Xanthoproteic Test

Amino acids containing an aromatic nucleus (tyrosine, tryptophan, phenylalanine) form yellow nitro derivatives on heating with conc. HNO_3. The salts of these derivatives are orange in colour. Proteins containing these amino acids also give a positive response to this test.

Reagents

1. Conc. HNO_3.
2. NaOH solution (40 per cent, w/v): Dissolve 40 g of NaOH in water and make the final volume to 100 ml.
3. Test solutions: Prepare separate solutions containing 50 µg/ml of amino acids like tyrosine, glycine, tryptophan, phenylalanine, lysine, cysteine, leucine etc.

Methodology

Add 1 ml conc. HNO_3 into 1 ml of test solution. Cool and then slowly pipette NaOH till the solution becomes alkaline. Note whether the mixture turns orange red in colour. Appearance of orange red colour denotes presence of aromatic amino acids.

Exp. 11: Detection of Proteins by Millon's Test

Amino acids or compounds containing hydroxybenzene radical react with Millon's reagent which constitutes mercurous and mercuric nitrates containing HNO_3 to form a red complex. Thus this test is specific for tyrosine.

Reaction

$$Hg + 4\,HNO_3 \rightarrow Hg(NO_3)_2 + 2NO_2 + 2H_2O$$

$$HO\!\bigcirc\!CH_2 - \underset{\underset{\text{Tyrosine}}{NH_2}}{CH} - COOH + Hg(NO_3)_2 \rightarrow \text{Complex product (red)}$$

Reagents

1. Millon's reagent: Dissolve 10 g of Hg metal in 20 ml of conc. HNO_3, dilute to 100 ml with water after cessation of evolution of the brown fumes.
2. Test solutions of different amino acids made as in the preceding test. Use a solution of phenol also as a test sample.

Methodology

Add 2-5 drops of Millon's reagent into 1 ml of test solution. Shake the contents and place the test tubes in boiling water bath.

Development of a red colour indicates presence of tyrosine. A positive reaction will also be obtained for proteins which contain tyrosine. Does phenol react positively to this test ?

The same experiment may be carried out with a modified Millon's reagent which contains 10 per cent mercuric sulphate in 10 per cent H_2SO_4. To 1 ml of test solution add 1 ml of modified Millon's reagent and boil gently for 1 min. Cool under running tap water and then add one or two drops of 1 per cent $NaNO_2$ solution and heat slightly. Appearance of red colour denotes presence of tyrosine. The advantage of using modified Millon's test lies in its being less susceptible to interference from inorganic salts.

Exp. 12: Detection of Proteins by Hopkins-Cole Test

The indole group of tryptophan reacts with glyoxylic acid in the presence of conc H_2SO_4 to give a purple colour. Glyoxylic acid is prepared by reducing oxalic acid with magnesium powder or sodium amalgam. Glacial acetic acid which has been exposed to the sunlight also contains glyoxylic acid and can thus be used for this test.

Reagents

1. Test reagent: To 10 g of magnesium powder add enough water to submerge it. Then slowly add 250 ml of cold saturated oxalic acid solution. Keep cool. Filter magnesium oxalate formed. Acidify the filtrate with 25 ml of glacial acetic acid and dilute it to 1 L with water. Alternatively, glacial acetic acid which has been kept in sunlight for a few minutes may be used as a test solution.
2. Test solutions: Prepare solution containg 50-100 μg/ml of individual amino acids such as tryptophan, tyrosine, phenlalanine, glutamic acid etc.
3. Conc. H_2SO_4.

Methodology

Add 1 ml of Hopkins-Cole reagent to 1 ml of test solution. Mix and pour with pipette conc. H_2SO_4 along the sides of test tube. Observe for appearance of blue or violet ring at the interface of the liquids which would indicate presence of tryptophan.

Exp. 13: Detection of Proteins by Biuret Test

This test is a general test for compounds having a peptide bond.

Alkaline copper sulphate reacts with compounds containing two or more peptide bonds to give a violet or pinkish coloured product which is due to formation of coordination complex of cupric ions with unshared electron pairs of peptide nitrogen and oxygen of water.

Reagents

1. Copper sulphate: 1 per cent (w/v) of $CuSO_4.5H_2O$ in water.
2. 40 per cent (w/v) solution of NaOH.
3. Test solutions:
   (*a*) Prepare 0.5 per cent (w/v) solution of a protein like casein or bovine serum albumin in NaOH and also a tripeptide glutathione.
   (*b*) Prepare 0.5 per cent (w/v) solutions of glutamate, cysteine and glycine.

Methodology

Add 0.5 ml NaOH to 1 ml of test solution and mix well. Then add 2-5 drops of copper sulphate and observe for the formation of pink or violet colour. Appearance of pink or violet colour shows presence of peptides or proteins in the sample. Try this test with solutions of glutathione and its constituent amino acids and record the observations.

Exp. 14: Determination of Crude Protein by Micro-kjeldahl's Method

The amino nitrogen in various nitrogenous compounds in biological samples such as plant materials (leaves, seeds etc.) is converted to ammonium sulphate on digestion with concentrated sulphuric acid in presence of K_2SO_4 and $CuSO_4$. K_2SO_4 is included in digestion mixture to raise the boiling temperature whereas $CuSO_4$ acts as a catalyst. Alternatively, solenium dioxide can be used as a catalyst. On distillation of the digested sample with NaOH, NH_3 is liberated and trapped in boric acid containing a mixture of bromocresol green and methyl red as an indicator. Ammonia reacts with boric acid to form ammonium borate which is then determined volumetrically by titrating against standardized HCl and the amount of nitrogen determined. Since proteins contain about 16 per cent nitrogen, the protein content of the sample is calculated by multiplying nitrogen content by 6.25. The value obtained is referred to as crude protein content since this procedure estimates total organic nitrogen rather than true proteins. This method is commonly used for determination of crude protein content of grains, forage and animal feeds.

Reagents

1. Conc. H_2SO_4.
2. Catalyst: Either selenium dioxide powder or a mixture of $CuSO_4 . 5H_2O$ and K_2SO_4 in the ratio of 2:1.
3. 40 per cent NaOH: Take 400 g of NaOH, dissolve it in water and make the volume to 1 L.
4. 2 per cent boric acid: Dissolve 10 g of reagent graded boric acid in about 470 ml of hot distilled water, cool and add 2 ml of 0.1 per cent alcoholic solution of bromocresol green and 4 ml of 0.1 per cent methyl red solution. Make the final volume to 500 ml.
5. 0.01 N HCl

Methodology

1. Take 1 g of finely powdered grain sample in a long necked digestion flask. Add 10 ml of conc H_2SO_4 and 200 mg of catalyst mixture and digest the sample on heater or Kjeldahl's heating unit till the solution becomes clear. Cool it and transfer to 50 ml volumetric flask. Rinse the digestion flask several times with small amount of water and pour the washings in the volumetric flask. Finally make up the volume to 50 ml with distilled water.
2. Take 10 ml of boric acid solution in 100 ml conical flask. Place this receiving flask in such a way that outlet of the condenser of micro-kjeldahl's distillation apparatus dips into the boric acid solution.
3. Transfer 5 ml of acid digested sample to the stream chamber of micro-kjeldahl's apparatus. Add about 5-8 ml of 40 per cent NaOH to the aliquot of the digested sample. Immediately close the stopcock and pass the steam through the steam chamber to distill ammonia till about 30-40 ml of distillate is collected in the receiving boric acid containing flask.

4. Remove the receiving flask, rinse the condenser outlet tip into the receiving flaks with water.
5. Titrate the contents of receiving flask against 0.01N HCl till the bluish green colour changes to pink.
6. Run a blank preparation which has been identically prepared except that it does not contain the sample.

Calculations

1 ml of 0.01 N HCl = 0.00014 g nitrogen

Since average nitrogen content of most proteins is 16 per cent

1 g of nitrogen = 100/16 g of protein

Volume of 0.01 N HCl used for blank = v ml

Volume of 0.01 N HCl used for sample = y ml

Titre volume of sample = y-v ml

Volume of sample taken for distillation = 5 ml

Total volume made of the digested sample = 50 ml

Nitrogen present in 5 ml of digestd sample = (y-v) 0.00014 g

$$\text{Nitrogen present in 50 ml digested sample} = \frac{(y\text{-}v) \quad 0.0014 \quad 50\text{ g}}{5}$$

$$\text{Nitrogen present in 1 g sample} = \frac{(y\text{-}v) \quad 0.00014 \quad 50\text{ g}}{5}$$

$$\text{Amount of nitrogen present in 100 g sample} = \frac{(y\text{-}v) \quad 0.00014 \quad 50 \quad 100\text{g}}{5}$$

$$\text{\% Protein content in sample} = \frac{6.25 \quad (y\text{-}v) \quad 0.00014 \quad 50 \quad 100}{5}$$

Exp. 15: Estimation of Protein by Lowry's Method

It is the most commonly used method for determination of proteins in cell free extracts because of its high sensitivity and quantities as low as 20 µg of proteins can be measured. The –CO-NH- (peptide bonds) in polypeptide chain react with copper sulphate in an alkaline medium to give a blue coloured complex. In addition, tyrosine and tryptophan residues of proteins cause reduction of the phosphomolybdate and phosphotungstate components of the Folin-cio-calteau reagent to give bluish products which contribute towards enhancing the sensitivity of this method. It is, however, important to remember that several compounds like EDTA, Tris, carbohydrates, NH_4^+, K^+, Mg^{++} ions, thiol reagents, phenols etc. interfere with the colour development and it should be ensured that such substances are not present in sample preparations.

Reagents

1. Phosphate buffer: (0.001M, pH 7.6) for preparation of the buffer.
2. Alkaline Na_2CO_3 reagent: Dissolve 2.0 g Na_2CO_3 in 0.1 N NaOH and make up the volume to 100 ml with 0.1 N NaOH.
3. Copper sulphate reagent: Prepare 0.5 per cent $CuSO_4.5H_2O$ in 1 per cent sodium potassium tartarate solution.
4. Alkaline copper sulphate reagent: Add 1 ml of Reagent iii to 50 ml of Reagent ii. This mixture is unstable and should be prepared fresh.
5. Folin's reagent: Dilute the reagent appropriately so that it is 1N in respect of its acid content.
6. 20 per cent (w/v) TCA: Dissolve 20 g of trichloroacetic acid in water and make the volume to 100 ml.
7. Acetone
8. 0.1 N NaOH
9. Bovine serum albumin (BSA): 100 mg/ml solution in distilled water.

Methodology

1. Sample extract: Weigh 1 g sample (germinating seeds, leaves, bacterial cells), macerate the sample in pestle mortar in 5 ml of phosphate buffer and transfer the material to centrifuge tubes. Centrifuge the homogenate at 8000 rpm for 20 min. Collect the supernatant and repeat the extraction 4-5 times. Combine the supernatants and make the volume to 50 ml with phosphate buffer.

2. Take 1 ml of the above extract and add 1 ml of 20 per cent TCA. Keep it for half an hour and centrifuge at 8000 rpm for 20 min. Wash the pellet with acetone twice and again centrifuge it. Discard the supernatant.
3. Dissolve the pellet in 5 ml of 0.1 N NaOH and mix well till it gets dissolved.
4. Take suitable aliquot (1 ml) of above solution and add to it 5 ml of reshly prepared alkaline copper sulphate reagent. Mix properly and after 10 min add 0.5 ml of Folin's reagent. Mix the contents instantaneously. Allow the colour to develop for 30 min.
5. Record the absorbance at 660 nm after setting the instrument with reagent blank which contains 1 ml of 0.1 N NaOH instead of the sample aliquot.
6. In another set of tubes take suitable aliquots of BSA solution (in a range of 0-100 µg) make the total volume to 1 ml with 0.1 N NaOH and develop the colour as described in steps 4 and 5. Draw a standard curve of absorbance at 660 nm versus µg of BSA. From this standard curve determine the amount of protein in the sample tube.
7. Calculate the amount of protein per g of the sample.

Precautions

1. It is important that contents of the tube be mixed immediately after the addition of Folin-cio-caleteau reagent. Proper colour development will be affected if this step is unduly delayed.
2. Compounds like Tris, EDTA, thiols like 2-mercaptoethanol, phenols and inorganic ions including K^+, NH_4^+ and Mg^{++} interfere in this procedure of protein estimation.

3.10 Detection and Estimation of Lipids

Exp.16: Extraction and Detection of Lipids

Isolation and purification of total lipids from body tissues is done by the methods of Folch *et al.* (1957).

Reagents

Chloroform–methanol mixture–2:1 by volume (v/v)

Methodology

Chloroform, methanol and water (containing 0.04 per cent $CaCl_2$, 0.034 per cent $MgCl_2$. 0.58 per cent NaCl or 0.74 per cent KCl) are mixed in a separating funnel by volume in the proportion of 8:4:3, respectively. The lower and upper phase are collected in separate glass bottles and named as solution A and solution B, respectively for further use.

The sample tissue is dried in oven for about 72 hrs till there is no further weight loss. The dried tissue is homogenized with chloroform-methanol mixture (2:1) and the final dilution is made to the required volume (1.0 ml/10 mg of body weight). Homogenate is filtered through a fat free Whatman filter paper No. 1. To the extract obtained is added 0.2 its volume of distilled water and mixed throughly with a glass rod. The rod is rinsed in the tube with minimal amount of solution A. The mixture is ultimately allowed to separate into two phases by centrifugation at 2400 r.p.m. for 20 minutes. The upper phase is removed completely with a pipette. The inside wall of tube is rinsed with solution B (1.5 ml for 10 ml extract) in such a way as not to disturb the lower phase. Rinsing is repeated twice and upper phase contents are maximally removed. The remaining rinsing solution and lower phase are then made into one phase by addition of methanol until the contents are clear. These contents are then transferred into pre-weight glass vials. The solvent is evaporated and glass-vials are weighed again to quantify the lipids.

Sudan Dye Test

Sudan are dyes that are insoluble in water but soluble in fat. Place several grains of the dye in a test tube containing water. Shake the tube. Note that the water does not get coloured. Now add several drops of an oil to the test tube and let it stand for some time. You will notice that the oil collects at the bottom and becomes deep orange-red.

Test for Glycerine

To a little solution add a few drops of copper sulphate solution and then a little sodium hydroxide. A clear blue solution is obtained.

Exp. 17: Lipolysis in Isolated Fat Cells by Hormones and Determination of Glycerol

Lipolysis, the breakdown of triglyceride to glycerol and free fatty acids, is stimulated by several hormones, including adrenalin, glucagon, and some pituitary peptides. This cascade effect is very similar to the hormonal stimulation of glycogenolysis. The rate of glycerol production thus monitors lipolysis directly. In the following experiment the rate of production of free fatty acids and glycerol in the isolated fat cells are measured in the presence and absence of the hormone adrenalin.

The samples are deproteinized with trichloracetic acid and the glycerol is assayed by measuring the extinction change at 340 nm when glycerol is oxidized by glycerol dehydrogenase.

Reagents

1. Trichloro acetic acid (100 g/l)
2. Diethyl ether
3. Glycine buffer (0.1 mol/l, pH 9.5)
4. NAD solution (6 mg/ml in glycine buffer prepared just before use.
5. Standard glycerol (0.25 mmol/l)
6. Glycerol dehydrogenase (dilute in glycine buffer to give an activity of 1 µmol min-1 and min-1).

Methodology

1. Remove 0.5 ml of the cell suspension for free fatty acid assay and transfer the remaining 0.5 ml to a glass centrifuge tube.
2. At the same time set up a blank containing 5 ml of the incubation medium and add 1 ml of trichloracetic acid (100 g/l) to all the tubes, mix thoroughly, and centrifuge to remove protein.
3. Transfer the supernatant to a glass stoppered tube, extract three times with 5 ml of diethyl ether, and assay the aqueous layer for glycerol.
4. Pipette 0.5 ml of the NAD solution in glycine buffer into a series of test tubes and add 0.4 ml of the extract.
5. Include a blank of 0.4 ml of water and a standard of 0.4 ml of glycerol. Start the reaction by adding 0.1 ml of glycerol dehydrogenase and incubate at 37°C for 1 h.
6. At the end of this time add 2 ml of glycine buffer and read the extinction at 340 nm. Calculate the glycerol released in

the test sample by comparison with the standard glycerol solution.

Exp. 18: The Determination of Free Fatty Acids

The sample is shaken with chloroform and buffered copper nitrate solution when the free fatty acids are removed from their complex with albumin and form copper-fatty acid soaps, which are extracted into the organic phase. The amount of copper in the chloroform is equivalent to the amount of free fatty acids in the sample and this is assayed by adding sodium diethyldithiocarbamate, a specific and very sensitive colorimetric reagent for the estimation of copper.

Reagents

1. Copper reagent (270 ml of 2 mmol/l) triethanolamine, 30 ml of 1 mol/l acetic acid, and 300 ml of 100 g/l cupric sulphate).
2. Sodium diethyldithiocarbamate (1 g/l in n-butanol prepared just before use).
3. Chloroform
4. Palmitic acid (5 mmol/l in $CHCl_3$).

Methodology

(a) Extraction

1. Add 2.5 ml of the copper reagent to 0.5 ml of each test solution in a glass stoppered tube and shake vigorously using a vortex mixer.
2. Add 5 ml of chloroform and shake vigorously by hand for 1 min.
3. Separate the layers by centrifugation when three phases are obtained: an upper aqueous layer, an interphase containing denatured protein, and a lower chloroform layer containing the Cu salt of the free fatty acid.
4. Remove the upper aqueous layer with a Pasteur pipette and discard. Break the interphase layer with a fresh pipette and carefully transfer the chloroform to a test tube.
5. It is very important at this stage to avoid contaminating the chloroform layer with the other two phases, which contain a high concentration of copper ions.

(b) Estimation

1. Take 3 ml of the chloroform, add 0.5 ml of the sodium diethyldithiocarbamate reagent, and read the extinction at 440 nm in a spectrophotometer or colorimeter.
2. Prepare a range of concentrations of the standard solution of palmitate to cover the range 0-5 mmol/l.
3. Remove 0.2 ml of each solution, transfer to a series of test tubes and evaporate to dryness on a boiling water bath in a fume chamber.
4. Add 0.5 ml of water to each tube and carry out the assay as set out above.

3.11 Quantitative Estimation of Nucleic Acids and their Preliminary Characterization

Tissue is homogenised at neutral pH and is centrifuged at low speed to obtain DNA fraction. This material is extracted with 2M NaCl which rupture DNA-protein bonds and bring DNA into solution. DNA is precipitated with ethanol, redissolved and reprecipitated. The final product is dissolved in buffer to give viscous solution. The protein molecules may be denatured with any detergent such as sodium dodecyl sulphate or may be removed by phenol. To separate DNA from RNA, latter is hydrolysed with ribonuclease, living DNA to be precipitated alone.

Exp. 19: Extraction of Total Nucleic Acids from Tissues

Nucleic acids are extracted by grinding the tissues with ice cold trichloroacetic acid followed by centrifugation to separate them from the acid soluble compounds (nucleosides and nucleotides etc.). The precipitated nucleic acids are then extracted with ethanol-ether to remove the interfering lipid material. RNA is selectively solubilized from the residue with alkali and DNA subsequently recovered. Extracted DNA and RNA are then quantitatively estimated by their colour reactions with diphenylamine and orcinol, respectively, as outlines in experiments.

Reagents

1. Trichloroacetic acid (10 per cent, w/v)
2. Ethanol: ether (2:1)

3. 0.5N NaOH
4. 5 per cent $HClO_4$

Methodology

1. Macerate 2.5 g fish tissue in 10 ml of ice cold 10 per cent trichloroacetic acid (TCA)
2. Centrifuge the homogenate at 3000 rpm for 10 min.
3. Discard the supernatant and suspend the pellet in 5 ml of cold 10 per cent TCA.
4. Centrifuge the suspension at 3000 rpm for 10 min, discard the supernatant and suspend the precipitate in 5 ml of ethanol: ether mixture.
5. Again centrifuge as in step 4 and suspend the precipitate in 5 ml of 0.5N NaOH. Mix well and keep it for 18 h at 37°C.
6. Centrifuge this alkali treated preparation at 3000 rpm for 10 min. Decant the supernatant, which contains RNA in the hydrolyzed form as well as proteins, in a test tube.
7. To the precipitate, which contains DNA and some proteins, add 1 ml of 5 per cent $HClO_4$ and after mixing the contents well, heat it in a boiling water bath for 1 h. Cool the preparation and centrifuge it at 3000 rpm for 10 min. Collect the supernatant, which contains deoxyribonucleotides released from hydrolysis of DNA, make up to a known volume with 5 per cent $HClO_4$ and using diphenylamine reaction estimate the amount of DNA in a suitable aliquot of this preparation.
8. To the supernatant obtained in step 6 (which contains hydrolyzed RNA), add an equal volume of 10 per cent TCA. Centrifuge at 3000 rpm for 10 min. Collect the supernatant. Make upto known volume with 5 per cent $HClPO_4$ and estimate RNA using orcinol method.

Exp. 20: Estimation of DNA by Diphenylamine Reaction

This is a general reaction given by deoxypentoses. The 2-deoxyribose of DNA, in the presence of acid, is converted to ω-hydroxylevulinic aldehyde which reacts with diphenylalamine to form a blue coloured complex with absorbance maxima at

600 nm. Compounds such as furfuryl alcohol and arabinal, which can be converted into ω-hydroxylevulinic aldehyde will also give this reaction. In DNA, since only deoxyribose of purine nucleotides is released, the value obtained represents half of the total deoxyribose in the sample.

Reagents

1. Standard DNA solution: Dissolve calf thymus DNA (100 µg/ml) in 1 N $HClO_4$ by heating at 70°C for 15 min. Make different dilutions of this stock solution ranging from 20-100 µg DNA/ml using 0.5 N $HClO_4$.
2. 1.6 per cent (w/v) acetaldehyde: Prepare by dissolving 1 ml of ice cold acetaldehyde in 50 ml of distilled water.
3. Diphenylamine solution: Dissolve 1.5 g of diphenylamine in 100 ml of glacial acetic acid and 1.5 ml of conc H_2SO_4.
4. Diphenylamine reagent: Prepare by mixing 0.5 ml of 1.6 per cent of acetaldehyde and 100 ml of diphenylamine solutions. This solution must be prepared fresh.

Methodology

1. Take 2.0 ml aliquot of the sample in which DNA has to be estimated a test tube.
2. In another set of test tubes, pipette 2.0 ml of standard DNA solution of different dilutions. In one of the test tubes take 2.0 ml of 0.5 N $HClO_4$ as a reagent blank.
3. Add 4.0 ml diphenylamine reagent to all the tubes, mix the contents properly and keep at room temperature in dark for 16-18 h or overnight.
4. Record the absorbance at 600 nm.
5. Draw a standard cuve of A_{600} vs DNA concentration. From absorbance of the sample, determine the amount of DNA in it. Express the results as mg of DNA/g fresh weight of tissue.

Exp. 21: Determination of RNA by Orcinol Method

This is a general method for estimation of pentoses. Acid hydrolysis of RNA releases ribose which in the presence of strong acid undergoes dehydration to yield furfural. Orcinol, in the presence of ferric chloride as a catalyst, reacts with furfural producing a green coloured compound with absorbance maxima at 665 nm. DNA gives a limited positive reaction with orcinol test.

Reagents

1. Colorimeter or spectrophotometer
2. Boiling water bath
3. 5 per cent $HClO_4$
4. Standard RNA solution: Dissolve yeast RNA (500 µg/ml) in 5 per cent $HClO_4$. Make different dilutions to obtain solutions containing 100–500 µg RNA/ml with 5 per cent $HClO_4$.
5. Orcinol reagent: Dissolve 100 mg of ferric chloride ($FeCl_3.6H_2O$) in 100 ml of conc HCl and then add 3.5 ml of 6 per cent solution of orcinol prepared in alcohol.

Methodology

1. Take 2.0 ml solution of each of the dilutions of RNA standard solutions, test sample and 2.0 ml of 5 per cent $HClO_4$, as a blank, in different test tubes.
2. Add 3.0 ml orcinol reagent to all the tubes and mix properly.
3. Keep the test tubes in a boiling water bath for 20 min.
4. After cooling them, measure the A_{665} against blank.
5. Plot a graph between A_{665} vs amount of RNA and from this standard curve determine the amount of RNA in the provided sample. Calculate the amount of RNA in 1 g fresh weight of the sample.

Exp. 22: Quantitative Determination of DNA and RNA by Spectrophotometric Method

Both DNA and RNA exhibit strong absorption of ultra-violet light due to presence of conjugated double bonds of the constituent purine and pyrimidine bases. These have a characteristics absorbance maxima at 260 nm which is linearly related with concentration of DNA or RNA in the solution upto an O.D. value of 2.0. This spectrophotometric method can also be employed for judging purity of DNA and RNA preparations. Proteins are usually the major contaminants in nucleic acid extracts and these have absorption maxima at 280 nm. The ratio of absorbance at 260 and 280 nm, hence, provides a rough idea about the extent of contaminants in the preparations. A ratio between 1.8-2.0 is indicative of fairly pure DNA and RNA preparations but values less than 1.8 signify presence of proteins as impurities.

Reagents

1. Saline sodium citrate (SSC) solution: Prepare 0.015M solution of sodium citrate (pH 7.0) and dissolve NaCl so that its final concentration in solution is 0.15 M.

Methodology

1. Switch 'ON' the spectrophotometer and allow it to warm up for about 10 min. Adjust the wavelength at 260 nm and put 'ON' the UV lamp.
2. Set the instrument at zero absorbance with SSC solution.
3. Read absorbance of solution of the provided sample. If O.D. is too high, appropriately dilute the sample solution with SSC and again take the reading.
4. Calculate concentration of DNA and RNA in the sample from following formulae:
   (*a*) For double stranded DNA:

   $$\begin{array}{c}\text{Concentration of DNA in}\\ \text{sample solution } (\mu g/ml)\end{array} = 50 \quad A_{260} \quad \text{Dilution factor}$$

   (*b*) For RNA:

   $$\begin{array}{c}\text{Concentration of RNA in}\\ \text{sample solution } (\mu g/ml)\end{array} = 40 \quad A_{260} \quad \text{Dilution factor}$$

Exercise No. 4

Blood

Blood is basically a transport tissue of the body which maintains the composition of interstitial fluid within optimum limits by simple diffusion of elements and compounds. It's own composition is the result of adding food by the process of digestion and assimilation. Blood being circulating fluid represent the main transport system within and between the organs of the body. It carries nutrients, gasses, hormones and metabolic wastes and further serve as an effective cooling system and maintain acid base balance of the body. The most important blood components are the blood cells and the plasma proteins, known for valuable enzymatic and immunity functions. Blood is also involved in respiratory and excretory activities of the body organs. Most important blood constituents of diagnostic value are the blood nutrients and the metabolic waste products such as urea, uric acid, creatine and respiratory pigments, the haemoglobin. Blood glucose, pH, the blood cells and the immunoglobulins are other important parameters which are measured as clinical tests for the diagnosis of a variety of pathogenic or physiological disorders. The immunoglobulins constitute major antibody system as an internal body defense system to control many antigens. Further most vital aspect of blood circulation is that it remain fluid and circulate freely within the vascular system. Blood loss due to any traumatic injury (haemorrhage) or rupturing of vessles is automatically controlled through a complex series of chemical reactions leading to the formation of blood clots (coagulation). The blood clotting mechanism is so organised that normally it do not operate so long blood is contained within the vessel due to the presence of anticoagulants like heparin. It however, quickly works at the place of injury and seal the leakage of the injured vessels. The blood coagulation parameters to be estimated are calcium, thermobin and fibrin contents of blood plasma.

The blood cells of teleost fish are produced in the hemotopoetic tissue, located in the spleen and kidney. There are no lymph modes. The leukocytes are divided into the lymphorcites, granulocytes and thrmbocytes. Thrmbocytes contain the blood clotting factors. The emthromcytes of fish are nucleated and similar in size to teh lymphocytes, in contrast to the mammals in which later on the much larger cell. The number of erythrocytes represent blood capacity for respiration due to its pigment (Haemoglobin) whereas the number of white blood cell determine the blood fighting tendency against pathogens. In pathological state the WBC count may be abnormally high (leukocytosis) or low (leukopenia). Infections are generally accompanied by leukocytosis. The platelets determine the capacity for blood clotting at the point of vessel injury. The blood cells in term differ greatly in composition.

The haemolysis of RBC in relation to RH factor is further important in blood physiology. The RH susbstance so called became it was first detected in *Rhesus* monkey and is present in 85 per cent of the white human population. The RH factor acts as an antigen. The human blood generally do not contain antibodies to RH factor, if a RH negative person receive RH positive blood, he develops antibodies against RH positive red cells which prove fatal.

The plasma proteins are generally as albumen, globulin and fibrinogen. The plasma proteins are easily precipitated by salting out process, *i.e.* 10 different proteins are precipitated with different salt saturation levels, such as fibrinogen is precipitated at 20 ppt.

The blood glucose values are obtained with capillary blood from a finger prick. In a normal healthy person, after feed, the blood glucose level rises from about 90 mg/100 ml to maximally upto 140 mg/100 ml. If venous blood is used, the blood glucose concentration is about 10 mg/100 ml during fasting and 25 mg/100 ml at the peak. A lower tolerance to glucose is found in diabetes mellitus due to lack of insulin, in diseases where the thryoid, pituitary or adrenals are overactive and also in severe liver disease when the rate of glycogen formation is reduced.

A higher tolerance to glucose is observed when the activity of the thyroid, adrenal or pituiary is decreased and also when absorption from the gut is impaired.

4.1 Collection of Blood

The blood can be collected by either Retro-Orbital breeding method *i.e.* from the orbital socket of mice or rat, or by severance of coudal peduncle in fishes or by cardiac puncture in all the animals. If blood is allowed to stand at room temperature for 2-3 h, then about half of the glucose is lost due to its conversion to lactic acid by glycolysis. Fluoride inhibits glycolysis, so blood is collected into an anticoagulant *i.e.* tube containing 1 mg of sodium fluoride for each millilitre of blood. Excess fluoride interferes with the glucose oxidase method, so it is prebably easier to collect 0.1 ml of blood directly into a blood pipette and transfer this immediately to the protein precipitant.

4.2 To Study Blood Groups and Different Types of Blood Cells

(A) Blood Groups

Human blood is generally classified into four groups. This grouping is based on the presence or absence of specific antigens found on the surface of the red blood cells. When an individual has only one of these antigens his blood is grouped as A or B type depending on the particular antigen present; when both are present the blood group is called AB; if both are absent the blood group is known to belong to O or zero group.

Antibodies, the substances complimentary to antigens, occur in blood plasma. The serum of the blood group A always contains the specific protein called beta (β); blood group B has the alpha (α) protein; type AB has neither; and type O carries both α and β proteins. The α protein is an antibody directed against the blood group susbtance A as a result of which if blood having RBC of A group is mixed with a antibodies in the blood plasma of A group blood, the latter, being anti B, would behave in a similar manner. This implies that the blood group A does not contain α protein and the group B lacks β protein, AB has neither but O type has both α and β proteins. Thus the antigen antibody composition of the four blood types is:

Aa, Ba, AB and O aB. The β protein will cause the clumping of the erythorcytes of the types B and AB; α protein will clump the types A and AB bloods in the antibody antigen reactions. Obviously, these bloods should not be mixed. The main consideration in the blood transfusion is to avoid the clumping of the donor's blood in the vessels of the recipient. The recipient's RBC do not usually clump because the donor's plasma antibodies are sufficiently diluted during transfusion. We, therefore, consider the effect of the serum of the recipient upon the corpuscles of the donor. Thus the individuals having the group AB are 'universal recipients' since they do not have α and β proteins in their blood and, therefore, do not produce any clumping of donor's blood cells. The individuals with blood group O are 'universal donors' since they have neither α nor β antigens and, therefore, whatever be the composition of the receiver's blood their RBC do not clump.

Transfusion of a particular blood type can be administered to the individual with the same type of blood, called homotypic, because they have identical antigen and antibodies.

Exp. 23: To Find Out Blood Group

For typing your own blood divide a slide in two halves with a glass marking pencil. Place a drop of anti A serum on the left side and a drop of anti B serum on the right side. Make an identification mark to indicate the sides on which sera A and B have been placed to avoid confusion. Prick the tip of your finger with a sterilized needle and transfer a drop of blood to the drop of anti A serum and stir it a little with the end of a tooth pick. Transfer another drop of blood to anti B serum and similarly mix it well. Cover the blood drops with cover slips. If the blood is in sufficient amounts a little of sodium citrate may be added to it. This will check coagulation of blood. Examine the slide under a microscope immediately, as delay may effect the coagulation which may be confused for clumping of RBC, and find out if you are a universal donor, universal recipient or none.

(B) Blood Cells

Blood in vertebrates is a highly complex fluid containing different types of inorganic and organic susbtances dissolved in the fluid, called blood plasma, and three types of cells, the red blood cells or erythrocytes, the white blood cells or leukocytes and the blood platelets. About fifty per cent by volume of blood in human beings makes blood plasma and the rest of it, making forty five per cent by volume, is composed of blood cells.

Exp. 24: To Prepare and Stain a Blood Film and Identify Different Types of Blood Cells

Blood film can be prepared either on a slide or on a cover slip. Clean the tip of your finger with spirit and allow it to dry. Prick the finger tip deep so that blood oozes freely from it. Touch one end of a slide with the oozing blood and collect some of it on the edge of the slide. Place the slide on a table and hold another slide between your thumb and index finger and touch its edge with the drop of the blood. The blood will spread along the edge of the slide. Hold this slide firmly at 45°C to the first slide which should be held firmly with thumb and the index finger. Now move the second slide forward, towards its other edge, with light and uniform pressure throughout. The blood drop will spread over the surface of the slide as a film. Allow it to dry.

Cover the dry film with Leishman's stain [Leishman's powder 0.15 gm, methyl alcohol (acetone free) 100 ml] for one minute. Now add an equal amount of buffer solution (disodium hydrogen phosphate 25 gms, potassium dihydrogen phosphate 32.5 gms and distilled water 100 ml, 2 ml of this solution should be added to 100 ml of distilled water) or distilled water and mix the two by gentle blowing. Leave it for ten minutes. Wash with tap water and allow it to dry. Examine the slide under high power or oil immersion lens and identify the following types of blood cells:

1. *Round cells*, without nucleus, with yellowish stain in the centre due to the change in refractivity and stained pink most abundant and evenly distributed, are the red blood corpuscles (RBC) or erythrocytes.
2. *Bigger cells*, relatively smaller in number, with granular or agranular cytoplasm and lobed or unlobed nuclei are the white blood corpuscles (WBC) or leukocytes, which take light blue stain but their nuclei become deep blue.
3. *Small dot-like corpuscles*, about one-third the size of the red blood cells and found lying in bunches among them, are the platelets or thrombocytes which take blue stain.

Leukocytes or WBC may have granular or agranular cytoplasm. Leukocytes with granular cytplasm are known as granulocytes. Granulocytes are of the following kinds:

1. *Polymorphys* or neutrophils make 55 to 70 per cent of the white blood cells. They take pink volet stain and are distributed irregularly in blood plasma. They have several nuclei which assume varying shapes and are usually connected by chromatin threads. The nuclei of the polymorphs have two to five lobes which may be located centrally or displaced towards the periphery.
2. *Eosinophils* are larger than the polymorphs and contain either a single irregular nucleus or two or three nuclei of unequal size. The cytoplasmic granules have strong affinity with acid dyes like eosine. They make 2 to 4 per cent of the total of the white blood cells.
3. *Basophils* are irregular in size and occur very rarely in the normal blood. They make less than 5 per cent of the total blood cells. Their nucleus is irregular and single. The granules of basophils are metachromatic.

Agranulocytes have non-granular cytoplasm. They have the following kinds:

1. *Monocytes* are the largest agranulocytes. Their nuclei are kidney or horse shoe-shaped. Their cytoplasm is stained clear blue or grey blue. They form only 1 per cent of the total white blood cells.
2. *Large lymphocytes* are little larger than the red blood cells. The nucleus is relatively large and round with very small extra-nuclear clear cytoplasm and stained blue with basic dyes. They make about 20-40 per cent of the total colourless corpuscles.
3. *Small lymphocytes* are very rare and make about two per cent of the total white blood cells. The cell body is similar to but smaller than large lymphocytes. The nuclei of small lymphocytes present gradations between oval and lobed conditions.

Discrimination between different types of white blood cells would appear difficult in the beginning but careful observation will make the task easier. Following points will help in distinguishing different types of WBC.

When stained with Leishman's dye the nucleus of leucocytes takes violet colour; the granules present in the extra-nuclear cytoplasm of neutrophils appear pink and fine in size; in eosinophils they are pink and coarse and in basophils they become blue and coarse.

If the stain is not properly prepared the above stated features will not be discernible. In such a case the distribution of granules will make a distinguishing index. In neutrophils blue and fine granules are distributed all over the surface; in eosinophils coarse granules are found aggregated on the surface of the nuclear membrane and in basophils coarse granules are distributed in the neighbourhood of but not on the nuclear membrane.

Study different types of blood cells. Draw the figures of each type and use different colour pencils to reproduce their stains.

Exp. 25: To Make a Differential Count of White Blood Cells

Determination of the precentage of differential types of white blood corpuscles is called the differential count. The differential count of the white blood cell is very important as it is a diagnostic character of many ailments. For example, polymorphs increase in number markedly in most acute infections; an increase in the eosinophils suggests the infection of foreign protein, asthama, allergy, or the prersence of chronic lead poisoning. An increase in monocytes, called monocytosis, indicates chronic parasitic infection and increase in lymphocytes, or lymphocytosis, diagnoses chronic infection like tuberculosis, whooping cough and chronic bronchitis, etc.

Prepare a slide of blood film and stain it. Identify different types of white cell. Draw a square on your note book and divide it into hundred small squares. Now identify the white blood cells and note down their initial letters, one in each of the small squares. Move the slide in different directions and observe hundred cells by this random sampling method. Count the total number of each type of cells and obtain your result in percentage units.

The microscope used for counting white blood cells should be kept in vertical position and must have a diaphragm, condenser and a concave mirror reflector. The concave reflector would be useful in differential count in that it can focus light at different points; the light reflected by the mirror should be condensed at certain points and not be allowed to diffuse. The WBC will appear as doubly refractile bodies under the microscope.

The RBC Counting

Several methods are used for counting the blood cells but the principle of their counting involves the dilution of a small volume of blood with a known volume of saline water or dilution fluid; since the saline fluid has an osmotic concentration similar to that of blood, the shape and size of blood cells do not suffer any change. A minute quantity of this mixed solution is now examined under the microscope and the number of corpuscles in a given area is counted.

The pipette used for diluting blood is known as *Thoma-Zeiss haemocytometer*. It consists of a graduated capillary tube with its one end blown into a bulb. The total capacity of the capillary tube is one hundredth of the bulb attached to it. The blood is sucked in the capillary tube with the help of a rubber tube attached to the free end of the bulb.

Blood is taken in the pipette upto, say, 0.5 mark; the dilution fluid* is afterward drawn up the stem to 101 mark. Since the fluid only in the bulb contains a dilute solution of the dilution fluid and the blood and fluid in the capillary tube onto the mark 1 does not mix with the blood, this makes a dilution of two hundred times to the blood.

The other apparatus, called *Neubauer's counting chamber*, consists of a thick glass slide with a thick 1 mm deep, transverse bar running along its centre. This bar has two vertical grooves running parallel to one another in the thickness of the slide and separating the bar from the rest of the slide along its two margins. A third bar running along the sagittal plane of the slide connects the two transverse bars to form the letter 'H'. On either side of the sagittal bar are nine big squares, called haemocytometer for RBC large squares, each of which is 1 sq. mm in area. The central square, also 1 sq. mm in area, is divided into twenty five squares, each of which is 1/25 sq. mm in urea. Each 'small square' is further divided into four parts and hence the area of the smallest square is 1/400 sq. mm.

Exp. 26: To Make RBC Count of Your Own Blood

Prick your finger tip with a sterilized needle and bring the tip of the*Dilution fluid, also called Hayem's fluid, has the following composition:

Mercuric chloride	0.5 gm
Sodium chloride	1.0 gm
Sodium sulphate	0.5 gm
Distilled water	200 ml

Mercuric chloride in the fluid acts as a corrosive sublimate and fixes the RBC while sodium chloride and sodium sulphate maintain isotonicity of the solution. Pipette near the drop of the oozing blood. Now tilt the pipette in vertical plane. The drop of the blood will rise and fill up the bore of the pipette due to capillary action. Remove the pipette from the sample of blood when the capillary blood reaches the 0.5 mark. The blood can also be sucked slowly upto the 0.5 mark. Wipe the tip of the pipette carefully because any blood left on the outer side of the stem may later on be sucked in with the dilution fluid and thus affect the composition of the fluid in the pipette. Now suck the diluting fluid upto 101 mark. Close the two ends of the pipette firmly between your thumb and fore-finger and shake it to so that the blood and the saline fluid are thoroughly mixed. Discard the clear fluid in the capillary part of the pipette. Now hold the pipette at 45° to the ruled surface of the slide and gently touch the pointed end of the pipette to it. The ruled surface will be covered with the fluid containing blood corpuscles by capillary action. Cover the slide with the special cover slip provided with the counting chamber. When the coverslip fits in the space provided, a space, which is $1/10^{th}$ of a mm in thickness, is left underneath. This fact has to be taken into account in calculating the total counts of RBC and WBC. Allow the slide to stand for a few minutes so that the corpuscles sink and settle down on the surface of the slide.

Calculations

Study the slide under high power magnification and count the number of red blood cells in at least five squares, each containing sixteen small squares.

It will be observed that many red blood cells either lie just above the lines of the squares or touch them. In such cases any two lines of the frame of the large square (generally those on the top and left side are chosen arbitrarily) joining each other at right angle should be considered and the blood cells touching them or lying above them should be taken into account. Likewise, the blood cells lying above or touching the other two lines (the lines on the left and bottom of the square) should be neglected. The cells touching two boundary lines are generally disregarded from the count.

The total number of small squares observed is 16 5 = 80

If X is the total number of blood cells observed in the 80 squares, the average number of blood cells present in each small square will be X/80.

$$\text{The area of each small square} = \frac{1}{20} \times \frac{1}{20} \text{ sq. mm.} = \frac{1}{400}$$

$$\text{The volume of each small square} = \frac{1}{20} \times \frac{1}{20} \times \frac{1}{10} = \frac{1}{4000} \text{ cu mm.}$$

1/4000 cu mm contains X/80 Cells,
1 cu mm will have $= \dfrac{X \times 4000}{80}$ cells

But the fluid has been diluted 200 times

$$\text{The cells present per cu mm in the Blood} = \frac{4000 \times 200}{80}$$

In practice it will be found that no calculation is needed; an addition of 0,000 to the total number of RBC present in the five squares will make the RBC count.

Precaution

Check that there are no foreign bodies or air bubbles and the cells are evenly distributed on the slide.

WBC Counting

The RBC of WBC counting is similar to that of RBC counting but WBC pipette has markings of 0.5, 1 and 11 and, therefore, the dilution of blood is twenty times only. The dilution fluids used in white blood cell counting, called Turch's fluid. It has the following composition:

Genetation violet or Methylene blue	1 ml
Glacial acetic acid	3 ml
Distilled water (to make)	100 ml

Glacial acetic acid haemolysis the RBC and the stain used is absorbed by WBC nuclei.

The four squares at the four corners of Nebauer's counting chamber are used for WBC counting. Each of these squares has an area of 1 sq. mm and is further divided into sixteen smaller squares.

Exp. 27: To Make WBC Counting of Your Blood

A film of dilute blood is made on the surface of the counting chamber, similar to that for RBC counting and the total number of WBC in all the four squares, called "Intermediate squares' is counted.

Calculations

Let X be the total number of white blood cells present in the four squares.

The average number of WBC present in 1 sq. mm area = X/4 *i.e.* X/4 WBC are present in X 1 X X 1/10 cu mm area.

Exp. 28: To Make Blood Platelet Counting of Your Blood

Blood platelets are round or oval, small, non-nucleated blood cells, about one-third the size of RBC. Their cytoplasms take faint blue colour with Leishman's stain.

Sodium citrate is mixed with the blood used for the slide preparation for platelet counting because it checks them from going into blood clotting.

Put a drop of 3 per cent sodium citrate solution on your finger tip and prick it through the solution. The oozing blood will mix with the solutions. Transfer the blood mixed with the solution to a clean slide and prepares a film. Dry the slide and stain it with Leishman's stain.

Study the slide under high power magnification and small stage aperture. Move the slide in different directions and each time count the total number of red blood cells and blood platelets under different microscopic fields.

Take twenty observations and enter them in three columns of your note book as follows:

No of field	No. of RBC	No. platelets
1	52	5
2	67	2

Suppose the total numbers of RBC and blood platelets in the twenty fields are X and Y, respectively.

With X RBC there are Y platelets.

With 1 RBC there will be Y/X platelets,

5,000,000 RBC* will have Y/X 5,000,000 platelets.

Precautions

Dust particles should not be mistaken for platelets- the platelets have clear boundaries whereas dust particles and stain granules have neither a clear boundary of cell membrane nor a distinct cell membrane.

4.3 Haemoglobin Content of Blood

The major mass of blood, as already stated, consists of haemoglobin present in red blood cells. The amount of haemoglobin present in blood is very important from the clinical point of view because it is the principal carrier of oxygen. The unit used for haemoglobin concentration in "100 per cent haemoglobin". The haemoglobin concentration for this unit varies with different authors. According to Haldane it is a blood that contains 13.8 gms/100 ml and the females have 13.7 gms/100 ml and is said to have cent per cent haemoglobin. Fifty per cent haemoglobin would thus be 13.8 12 or 6.9 gms haemoglobin per 100 ml. For Sahil 100 per cent haemoglobin is equal to 17.3 gms per 100 ml.

On full saturation one gram of haemoglobin carries 1.34 ml of oxygen; 13.8 gms of haemoglobin present per 100 ml of cent per cent haemoglobin blood can combine 13.8 1.34 or 18.5 ml of oxygen. This is called oxygen capacity of blood and depends on the haemoglobin content per 100 ml of the blood.

---

* An average number RBC per cubic millimetre is about 5,000,000 in human beings. The blood platelet count is expressed as per cubic millimetre.

Two factors, namely, the red blood cell count and the haemoglobin colour index, are taken into account for determining the amount of haemoglobin per cell of a blood sample:

$$\text{Colour Index} = \frac{\text{Amount of Haemoglobin in \%}}{\text{Number of RBC in \%}}$$

Thus, if there are 5,000,000 red blood cells per cubic millimetre (*i.e.*, there are 100 per cent blood cells) and 13.8 gms of haemoglobin per 100 ml of blood (*i.e.*, 100 per cent haemoglobin), the colour index is said to be one.

The most important factor affecting the haemoglobin content of blood is dearth of iron in food, which causes a lower haemoglobin percentage; other reasons may be physiological, such as deficiency in corpuscle formation, as in pernicious anaemia.

The most common and easy method for haemoglobin estimation by the physicians is by the examination of cornea and nail beds. Accurate estimation of haemoglobin is made in the laboratories by photo-electric colorimeter; reasonably accurate assessment of haemoglobin contents can be made by Halen's and Sahil's methods which are most popular methods for the purpose. These methods are based on the principle of matching the colour of the blood to be tested with that of the standard blood. The apparatus used for haemoglobin estimation is known as haemoglobinometer. It consists of three tubes, two of which, called standard tubes, contain 20 cu mm of standard blood (*i.e.*, the blood capable of carrying 18.5 ml per cent of oxygen, or containing 13.8 gms of haemoglobin per 100 ml) each, diluted to 200 mark, or these are replaced by glass tubes coloured to the shade of the standard blood. The third tube, called diluting tube, is inserted between the standard tubes and used for the comparison of its blood colour with those of the standard tubes.

Exp. 29: To Find Out Haemoglobin Content of Blood

Fill the diluting tube with 20 cu mm of N/10 HCl (9 ml of HCl in 1000 ml of distilled water). Now prick the tip of your finger and such 20 cu mm of blood in a pipette. Transfer this blood by blowing the pipette gently in diluting tube. Rinse the pipette several times so that no blood is left in it. Make sure that air bubbles are not present in the diltuing tube. Allow the tube to stand for ten minutes so that the haemoglobin is acted upon by the hydrochloric acid and converted into acid haematin. Now dilute the blood further, drop by drop, with distilled water or N/10 HCl and stir the solution constantly with a glass rod till the colour of the blood in the diluting tube matches with that of the standard tubes. Dilution of blood is read in terms of gram per cent; for example, if the colours of the blood in the diluting and standard tubes are alike when the dilution is 50, the blood is said to contain only 50 per cent of the normal haemoglobin.

4.4 Haemin crystals

When an acid or alkali is added to haemoglobin, it breaks up into a brown pigment, called haemtin, and a protein, called globin. The hydroxyl group of haematin is later on replaced by chlorine to produce haemin, which separates out as dark brown prisms and star-shaped clusters with rounded edges. Haemin test distinguishes blood from red ink and beatle stains, etc. and, therefore, it has a great medico-legal important. In case of an old blood stain sodium chloride is added to haematin for the production of haemin but fresh blood does not require it because it contains enough sodium chloride in its plasma.

4.5 Effects of Isotonic, Hypertonic and Hypotonic Salt Solutions on Red Blood Cells

The pressure exerted by the contents of red blood cells on the cell membranes is the same as that of the blood plasma, that is, a condition of osmotic equilibrium exists between the two. It means that the rate of entry of water molecules into the blood cells is the same as that of those passing out of them. Normal human blood plasma exerts a pressure equal to that of 0.9 per cent solution of sodium chloride. It is known as isotonic solution. When the red blood cells are kept in an isotonic solution they remain unaffected. If the concentration of the solutes in the fluid outside a cell is lower than 0.9 per cent sodium chloride solution, the solution is known as hypotonic saline solution. Hypertonic saline solution contains higher than 0.9 per cent of sodium chloride in it.

If water is added to blood plasma it becomes diluted; it results in an immediate fall in the osmotic pressure of the blood plasma, but the osmotic pressure inside the cells remains unchanged. Consequently, the water from the plasma will start entering the cells to equalise the osmotic pressure within and without the cells. The blood cells will thus change from bioconcave to biconvex outline, swell into spheres and ultimately burst, discharging their haemoglobin in the surrounding fluid. The release of haemoglobin due to rupture of cell membrane of the red blood cells is known as haemolysis. If the red blood cells are kept in a hypertonic saline solution, where the concentration of the solute molecules is greater than that of the blood cells, the direction of the flow of water is from the cells into the blood plasma. This results in shrivelling of the blood cells due to the loss of water.

Exp. 30: To Study the Effects of Isotonic, Hypotonic and Hypertonic Solutions on Red Blood Cells

Take three glass slides and place a drop of isotonic solution (0.9 per cent salt solution) on one, a drop of hypertonic solution (2 per cent solution of sugar) on the second, and a drop of hypotonic solution (0.4 per cent salt solution) on the third. Add a drop of blood to the solution on each slide; place coverslips and observe the effects of different solutions on the blood cells under the high power objective.

4.6 Blood Coagulation

Coagulation or solidification of blood is brought about by a series of interlinked chemical reactions. The formation of blood clot or thrombus involves the conversion of a specific plasma protein, called fibrinogen, from its sol to the gel fibrin state. The process of clotting starts by the action of a substance, thromboplastinogenase, released from blood platelets and to a lesser extent from the damanged tissue cells. Thromboplastinogenase and certain plasma globulin components, called anti-haemophilic factors, act on thromboplastinogen, an inactive globulin component of blood plasma, and convert it into an active state of a lipoprotein called thromboplastin. Thromoplastin, so produced, and calcium ions, which are normally present in the blood, act together and this converts the former into thrombin. Thrombin finally acts on fibrinogen and converts it to fibrin.

Exp. 31 To Determine the Coagulation Time of Your Blood

For finding out the coagulation time fine glass capillary tubes are used. Prick your finger tip with a needle and bring the end of the capillary tube in contact of the blood drop. The blood will be sucked up in the tube by capillary force. Note the time of the sucking up of the blood by the capillary tube with the help of a stop watch. Keep the tube in a horizontal plane and observe the movement of blood at an interval of every 30 seconds. When the blood movement stops break the capillary tube in small pieces every fifteenth seconds after scratching it with a file. The liquid blood on coagulation will be transformed into a fibrin thread. Note the time when fibrin thread appears in the capillary tube. The interval between the oozing of the blood from the finger tip and the appearance of fibrin thread is the coagulation time.

4.7 Analysis of Blood Metabolites

The blood plasma is deprotenized and the metabolites assayed in the supernatant after centrifugation.

Exp. 32: Blood Glucose and Protein Estimation

Blood glucose is estimated as stated in the experiment 4 and blood proteins following the method given in experiment 15.

Exp. 33: Blood Urea Estimation

Urea is the end product of protein metabolism and is produced in the urea cycle, arginase is the key enzyme in this cycle. Arginase catalyses the hydrolysis of arginine to ornithine and urea. The estimation of arginase activity is made by measuring the amount of urea released in the reaction.

Reagents

1. Blood plasma from the test fish (four fed and four starved for three days).
2. Perchloric acid (0.5 mol/l)
3. Bromophenol blue indicator solution.
4. KOH (1 mol/l)
5. –isonitrosopropiophenone reagent for urea estimation (2 g/l B-isonitrosopropiophenone containing 170 ml of conc. Sulphuric acid and 40 ml of syrupy phosphoric acid made up to 1 l with water)
6. Standard urea solution (0.5 mmol/l)

Methodology

Add 4 ml of perchloric acid to 1 ml of blood plasma, mix thoroughly, and remove the percipitate by centrifugation. Pipette 3 ml of the supernatant solution into a calibrated centrifuge tube, add one drop of bromophenol blue indicator solution and KOH until the colour changes from yellow to blue. Add distilled water to the 9 ml mark, mix thoroughly, and leave in an ice bath for 30 min. At the end of this time remove the precipitate of $KClO_4$ by centrifugation and store on ice if used the same day or frozen if used in the next practical session.

Take 1 ml of the deprotenized plasma and determine the urea content as follows:

Mix 1 ml of sample with 6 ml of the -isonitrosopropiophenone reagent and heat in a boiling water bath with a marble on top. The red colour produced is sensitive to light, so all the test tubes are wrapped in aluminium foil. After 1 h, remove the tubes, transfer them to a beaker of cold water, and cover with a black cloth. After allowing the tubes to cool for 15 min, read the extinction at 520 nm. Prepare a standard curve with the 0.5 mmol/l urea and calculate the micromoles of urea released per minute. Express the final result as micromoles per minute per gramme of liver and the total activity present in the liver. The overall dilution of the plasma is x15 and this must be allowed for in the calculation.

Exp. 34: Determination of Free Fatty Acids

The sample is shaken with chloroform and buffered copper nitrate solution when the free fatty acids are removed from their complex with albumin and form copper-fatty acid soaps, which are extracted into the organic phase. The amount of copper in the chloroform is equivalent to the amount of free fatty acids in the sample.

Reagents

1. Copper reagent (270 ml of 2 mmol/l triethanolamine, 30 ml of 1 mol/l acetic acid, and 300 ml of 100 g/l cupric sulphate).
2. Sodium diethyldithiocarbamate (1g/l in n-butanol prepared just before use).
3. Chloroform
4. Palmitic acid (5 mmol/l in $CHCl_3$)

Methodology

Extraction

Add 2.5 ml of the copper reagent to 0.5 ml of each test solution in a glass-stoppered tube and shake vigorously using a vortex mixer. Add 5 ml of chloroform and shake vigorously by hand for 1 min. Separate the layers by centrifugation when three phases are obtained: an upper aqueous layer, an interphase containing the Cu salt of the free fatty acid. Remove the upper aqueous layer with a Pasteur pipette and discard. Break the interphase layer with a fresh pipette and carefully transfer the chloroform to a test tube. It is very important at this stage to avoid contaminating the chloroform layer with the other two phases, which contain a high concentration of copper ions.

Estimation

Take 3 ml of the chloroform, add 0.5 ml of the sodium diethyldithiocarbamate reagent, and read the extinction at 440 nm in a spectrophotometer or colorimeter.

Standard Curve

Prepare a range of concentrations of the standard solution of palmitate to cover the range 0-5 mmol/l. Remove 0.2 ml of each solution, transfer to a series of test tubes and evaporate to dryness on a boiling waterbath in a fume chamber. Add 0.5 ml of water to each tube and carry out the assay as set out above.

Exp. 35: The Effect of Detergents and Other Membrane Active Reagents on the Erythrocyte Membrane

Many detergents disrupt membranes by 'dissolving' the phospholipid components. In the case of erythrocytes, this effect can be readily followed in a colorimeter by measuring the absorbance of haemoglobin released from the disrupted cells.

Reagents

1. Isotonic saline (8.9 g/l)
2. Detergents (10 g/l). neutral, Triton X-100; cationic, cetyltrimethylammonium bromide; anionic, sodium dodecyl sulphate
3. Lysophosphatidyl choline (lysolecithin, 10 mmol/l)
4. Progesterone (100 mmol/l in ethanol)

5. Hydrocortisone (saturated solution in ethanol, approx 5 mmol/l)

Methodology

1. Use human blood from a transfusion bottle or collect the blood from a freshly killed rat into a tube containing 0.5 ml of an anticoagulant (40 g/l trisodium citrate).
2. Centrifuge the cells, wash them twice with isotonic saline and resuspend them in the same volume as the original blood.
3. Dilute the erythrocyte suspension with saline so that when 0.5 ml is added to 4.5 ml of Triton X-100, an extinction of about 0.8-0.9 is obtained at 540 nm after centrifugation. This represents 100 per cent lysis and all subsequent extinction values should be expressed as per cent lysis.
4. Pipette duplicate 0.5 ml samples of the diluted erythrocytes into 10 ml centrifuge tubes containing 4.5 ml of saline.
5. Mix thoroughly and add 50 µl of the test compound, mix by gentle swirling, and place in an incubator at 37°C for 20 min.
6. Separate any unbroken cells by centrifugation on a bench centrifuge and measure the absorbance of the supernatant solution at 540 nm.
7. First, examine a range of concentrations of the three detergents and prepare a graph of per cent haemolysis against concentration of the reagent.
8. Repeat the experiment with lysolecithin, progesterone, and hydrocortisone and comment on the relative lytic effect of these compounds.
9. Finally, examine the protective effect of hydrocortisone by incubating the erythrocytes with this steroid before exposing them to the lytic agent.

Exp. 36: The Separation of Serum Protein by Electrophoresis on Cellulose Acetate

There are numerous applications of electrophoresis in medicine and clinical biochemistry and one test frequently carried out is the analysis of serum for changes in the proteins during disease. A rapid and convenient method for this is cellulose acetate electrophoresis illustrated in the following experiment:

Reagents

1. Barbitone buffer (0.07 mol/l, pH 8.6)
2. Ponceau S protein stain (2g/l in 30 g/l TCA)
3. Acetic acid (5 per cent v/v)
4. Serum
5. Cellulose acetate strips (Oxoid Ltd)
6. Methanol: water (3:2)
7. Citrate buffer (0.1 mol/l, pH 6.8)

Methodology

1. Moisten a strip of cellulose acetate (10 cm 2.5 cm) by placing it on the surface of the buffer in a flat dish and allow the buffer to soak up from below.
2. Immerse the strip completely by gently rocking the dish, then remove it with forceps.
3. Lightly blot the strip, place in the apparatus, and connect it to the buffer compartments with filter paper wicks.
4. Switch on the current and adjust to 0.4 mA per centimetre width of strip.
5. Apply a streak of serum from a microlitre pipette or melting point tube about one-third of the length of the strip from the cathode.
6. Carry out electrophoresis for 11/2 -2h, remove the strips, and stain with Ponceau S for 10 min. Prior heating is not needed here since the TCA fixes the proteins for staining.
7. Remove excess dye from the strip by washing repeatedly in 5 per cent v/v acetic acid.

Repeat the experiment after precipitating some of the proteins with ice-cold methanol. All steps must be carried out at 0°C. Mix 2 ml of plasma with 1 ml of the citrate buffer in a centrifuge tube. Slowly add 7 ml of ice-cold methanol in water with stirring. Leave on ice for 30 min, then centrifuge at 3000 g for 10 min in a refrigerated centrifuge. Remove the supernatant and dissolve the precipitate in

1-2 ml of saline. Carry out electrophoresis of the dissolved precipitate and also the supernatant and discuss the results.

Exp. 37: The Estimation of Blood Cholesterol by Liebermann-Burchard Method

This test is based on the fact that the steroids react with sulphuric acid in the presence of acetic anhydride to produce a characteristic blue green colour. Blood or serum is extracted with an alcohol-acetone mixture which removes cholesterol and other lipids and precipitates protein. The organic solvent is removed by evaporation on a boiling water bath and the dry residue is dissolved in chloroform. The cholesterol is then determined colorimetrically using the Liebermann-Burchard reaction. It is essential to use absolutely dry glassware for this estimation.

Reagents

1. Alcohol–acetone mixture (1:1)
2. Chloroform
3. Acetic anhydride–sulphuric acid mixture (80:1 mix just before use)
4. Stock cholesterol solution (100 μg/ml in chloroform)
5. Working cholesterol solution. (Dilute the above solution one in five with chloroform to give a solution of 0.4 mg/ml).

Methodology

1. Place 10 ml of the alcohol-acetone solvent in a centrifuge tube and add 0.2 ml of serum or blood.
2. Immerse the tube in a boiling water bath with shaking until the solvent begins to boil
3. Remove the tube and continue shaking the mixture for a further 5 min.
4. Cool to room temperature and centrifuge
5. Decant the supernatant fluid into a test tube and evaporate to dryness on a boiling water bath.
6. Cool and dissolve the residue in 2 ml of chloroform. At the same time, set up a series of standard tubes containing cholesterol and a blank with 2 ml of chloroform.

7. Add 2 ml of acetic anhydride-sulphuric acid mixture to all tubes and thoroughly mix.
8. Leave the tubes in the dark at room temperature and read the extinction at 680 nm.

Exp. 38: The Preparation of Cholesterol from Brain

Cholesterol is readily soluble in acetone, while most complex lipids are insoluble in this solvent.

Materials

1. Brains from calf or pig.
2. Acetone
3. Ethanol

Methodology

1. Add 400 ml of acetone to 100 g of brain and blend for 1 min.
2. Rinse out the blendor with a little acetone and stir the combined homogenates for 10 min.
3. Filter the suspension on a Buchner funnel, blend the residue with a further 200 ml of acetone as before, filter and combine the filterates.
4. After removing most of the acetone by distillation under reduced pressure, cool the flask with tap water and collect the crude cholesterol on a Buchner funnel.
5. Dissolve the crude material in the minimum volume of hot ethanol and filter while hot using a fluted filter paper. For this, the collecting flask is placed in a boiling water bath with the filter funnel in position.
6. Air dry the cholesterol and record the yield

Exp. 39: Study of Relative Plasma Lipid Peroxidation

Lipid peroxidation was performed on the blood plasma samples as to assess the total levels of lipid peroxidation in the blood. The Thiobarbituric Acid Assay (TABRAS) Test was employed to assess the levels of lipid peroxidation using Malondialdehyde, the byproduct of lipid peroxidation as a result of break down of polyunsaturated fatty acids (PUFAs), as the index of lipid peroxidation. As we cannot measure the actual concentration of the free radicals themselves in the blood plasma because of their very small lifetime after being produced, we measure the biproducts formed as a result of the production of free radicals and get a relative idea of the levels of free radicals produced. TBARS test is the most common and widely used test to assess the damage done by free radicals, as it is most simple and easy method. The working principle behind this test is that Melondialdehyde reacts with thiobarbituric acid to give a red species absorbing at 535 nm.

Reagents

1. 155 w/v Trichloroacetic acid
2. 357 per cent w/v Thiobarbituric acid
3. 0.25 N Hydrochloric acid.

Procedure

1 ml of biological sample with 2 ml of stock (TCA-TBA-HCL) solution is mixed thoroughly. The solution is warmed at 80 degree centigrade for 20 minutes in water bath. After that the solution is cooled and the flocculent precipitates are removed by centrifugation at 1000 g for 10 min. The absorbance of the sample is observed at 535 nm against the blank that contains all the reagents except the lipid (biological sample). The more the absorbance, it indicates more the concentration of malondialdehyde which directly further indicates the extent of lipid peroxidation in the sample.

Exercise No. 5
Hormones

Various vital activities of the animal body are coordinated and controlled by the systems of nerves and endocine glands. The endocrine glands, also called the ductless glands, differ from other glands in that they do not possess any ducts to carry their secretions and, therefore, release the products directly into the blood, which carries them to different parts of the body. The endocrine system of glands brings about a chemical coordination of the body functions, such as growth, development, reproduction, energy utilization and cell permeability, by various chemical substances, called hormones, which either stimulate or inhibit a function but generally do not initiate a process. The hormones are effective in remarkably small amounts. They produce rather diffused effects, called secondary effects, on all the body cells which result from the transport of hormones round the body by the blood, but their definite action is produced on a specific target organ in the body. Since they are transported in the blood, the response produced by them is much slower but more prolonged than that produced by the nervous system. Many hormones have been identified in man and other animals; these include thyroxine, insulin, testosteron, estrogen, adrenalin, glucagon, melanocyte-stimulating hormone etc.

Exp. 40: To Demonstrate the Effect of Pituitary Extract on Frog/Fish

One or the hormones produced by the anterior lobe of the pituitary gland is gonadotrophic, *i.e.*, it stimulates the female frog to ovulate. In the month of October the extract of only one pituitary gland is enough to induce a female frog or fish to ovulate but as the time passes the dosage increases; it requires the extract of six pituitary glands to ovulate a female frog between the months of November and January; three glands are enough for the purpose in March and April. The extract of the anterior pituitary gland of the female frogs has been found to have double sex-stimulating capacity as compared to that of the males.

The animals used for collecting the pituitary extract should preferably be freshly caught. Take seven female frogs and keep the largest of them for the purpose of ovulation and use the other six females for the removal of the pituitary glands. First insert a pair of scissors in the jaw angle of one of the living frogs, cutting along the postero-median plane behind the eyes; quickly cut across the other jaw. Cut the upper palate of the skull lateral to brain on either side by inserting the scissors into the posterior brain cavity. These cuts should extend as far as the orbits. Turn back the upper palate and hold it in the position with your left thumb. A small oval, pink body, called pituitary gland, will be seen attached to the optic chiasma and surrounded by a white endolymphatic tissue. Remove the gland with a pair of forcepts and put it in a watch glass containing about 2 ml of water. Place similarly removed glands from the other five frogs in the same watch glass. Collect the contents of all the glands in the barrel of a hypodermic syringe and inject it into the abdominal lymph space of the seventh frog kept for the experiment. Pituitary extract purchased from drug stores can also be used for the purpose if sufficient number of frogs are not available. While injecting the extract care should be taken that the cutaneous and the anterior abdominal veins and other parts of the body are not injured. This is possible if the needle is inserted into the posterolateral part of the body. Put the injected female in a jar containing about 2.5 cm water in it.

The period for ovulation of the injected frog is controlled by the temperature at which it is kept. At 25°C, which is considered best, the female takes three days for ovulation. The female should be stripped gently 36 hours after the injection of the pituitarty extract. It is performed by bending the abdomen of the female at right angle to the hind limbs at the level of pelvis. Now apply pressure to both the sides of the abdomen so as to squeeze it gently in the direction of cloaca; this presses out the eggs from the oviducts through the cloacal aperture. Collect the eggs in 1 per cent Rigger's solution in the biologically clean glass dishes.

Exp. 41: To Study the Effect of Thyroxine on the Frog Tadpoles

Thryoid glands, besides other functions, also effect coordination in general growth and development of all the cells and tissues. Tadpoles required for this experiment can be collected from ponds in their very early stage of development, or they can be brought up in the laboratory by artificial fertilization of the eggs obtained by induced ovulation in the previous experiment; or from the fertilized eggs collected from pond and reared in the laboratory. Instructions on artificial fertilization of the eggs and their rearing in the laboratory are as follows:

Cut open the viscera of two male frogs, remove the pair of testes and macerate them in 20 ml of pond water. Keep the sperm suspension, so prepared, in this state for about 20 minutes. Pipette the sperm suspension over the eggs and shake the dishes to ensure an even distribution of sperms. After half an hour the fertilized eggs would be seen rotating within their jelly covers with their black animal poles. After half an hour, when the jelly covering would be completely swollen, cut the eggs clusters apart and separate the eggs from one another.

The fertilized eggs can also be collected from ponds and ditches in the month of mid-July and August. They appear on the surface of water as bunches of beads covered with a jelly-like susbtance.

Locate the fertilized eggs in any pond. Collect some water in a large bucket from the pond; put a few plants from the pond in the bucket, if available. Now introduce a dipper into the water in the close vicinity of the eggs and lift them with some water in such as manner that the position of the eggs is not disturbed. Transfer the eggs from the dipper in the bucket carefully by lowering the dipper into in until the egg mass floats free from the dipper. This will avoid the churning effect of the cytoplasm of the eggs during their transfer. Now remove the dipper from the bucket gently without disturbing the eggs. The eggs should in no case be transferred by pouring them in the bucket.

The eggs should be kept in a sufficiently large volume of water; about three gallons of water is enough for an egg mass of the size of a hen's egg. If a slowly running water is provided the possibility of successful rearing would be greater as it will drain away the metabolites discharged by the developing embryos. If the eggs are transferred from the dipper in the aquarium in the laboratory make sure that the temperatures of the pond water and that of the aquarium are not markedly different, otherwise let the dipper stand for some time until the temperature of the water in the dipper becomes equal to that of the laboratory.

Examaine the eggs with a hand lens. The development is fast at higher temperatures and can be slowed down, if desired, by placing the aquarium in a cooler place.

The hormone thyroxine situmulates the metamorphosis of frog from tadpole into an adult. The hormone is absorbed directly through the skin of tadpoles if they are kept in a solution of thyroxine and water. Put the tadpoles in:

1. Pond water without thyroxine
2. Solution with one part of thyroxine to five million parts of water, and
3. Solution with one part of thyroxine to ten million parts of water.

Prepare a solution of 10 mg of crystalline thyroxine in 5 ml of 1 per cent sodium hydroxide solution. Add distilled water to bring up its volume to one litre. This makes a stock solution with 1:100,000 concentration. Refrigerate it until ready for use. Dilute the stock solution five and ten times to prepare the solutions used in the experiment.

Feed the tadpoles on hard-boiled egg-yolk on alternate days. Change the water every week-end to avoid fouling of the medium.

The developing tadpoles can also be kept on the thyroxine-wheat flour mixture. The food can be prepared by crushing and dissolving 5 two grain thyroxine tablets in 5 ml of distilled water in a mortar. Add an equal weight of whole-wheat flour to it and grind up the mixture. Leave this paste spread in a thin layer on a glass sheet for drying. When dried, powder the dry mixture and store it in closed bottles in a refrigerator. Use 50 mg of thyroxine-wheat powder per tadpole per day, daily for a week. Feed the tadpoles, both controlled and uncontrolled, on par-boiled spinach or lettuce also.

Change the water daily to avoid fouling. Record the changes in the length of tail and hind-limbs, time of appearance of fore-limbs, changes in the shape of the head and body. Compare your observation with those of the tadpoles that have not been given thyroxine.

Exp. 42: To Study the Effect of Thyrodectomy on Fish

The thyroid gland consists of two elongated oval lobes, one on either side to a trachea, joined near their posterior poles by a thin isthmus crossing the trachea ventrally. The lobes, buried under a membrane of the neck region, are richly vasularized and made up of groups of hollow spheres visible microscopically. The lobes extend as far as cricoid cartilage of the larynx anteriorly and over the first three of four tracheal rings on the posterior side in the specimens weighing more than 100 gms but their size is smaller in the specimens which are less than 100 gms in weight.

For the experiment on thyrodectomy large sized fishes should be selected because the gland is easily located in them. Find out the weight of the fish by weighing it. Now hold the fish,wrapped in a piece of cloth, in your left hand in such a way that its ventral side faces you and inject any anaesthesia, say neumatol, at a rate of 30 mgms per kg, with a sterilized hypodermic syringe. The animal can also be anaesthesized by putting it in a jar containing ether (anaesthetic) vapours.

Now lay the animal in a dissection tray and given an incision along its mid-ventral line in the neck region with sterilized instruments. Open the incision wide, remove the membrane covering the cricoid cartilage. You will observe a dark brone band, about 2 mm in length, along the lateral margins of the cartilage projecting on either side. These are the lobes of the thyroid gland. Give one cut on the anterior end and another on the posterior end of the gland with a pair of sharp scissors. Take care that blood vessels are not injured during the operation. The bleeding, if any, can be checked by covering the cut end of the vessel with a cotton gauge soaked in adrenalin diluted in water (1:1,000) or immersing the tissue in warm normal saline, which acts as vaso-constrictors and, therefore, check bleeding.

Now stitch the incision with cat's gut no. 4/0 with a needle, first stitch the lower muscle layer and then the outer skin. Apply some antiseptic, detol or savlon, on the stitched incision line, cover it with a cotton pad and dust a little cibazol powder. Seal it with tincture benzoin; this will check infection and pus formation.

Transfer the animal to its cage and observe the effects of thyrodectomy. Following changes will be observed in the fish. The general behaviour of the animal becomes sluggish and slow; its appetite goes down; swellings appear all over the skin and the fish starts gaining weight. Keep a record of the increase in weight every week. The general metabolism and oxygen consumption become less. Compare the rate of oxygen consumption of a fish without thyroid gland with that of a normal fish.

Exercise No. 6

Enzymes

The enzymes are basically the proteins produced by the cells which catalyse specific chemical reaction such as oxidation, reduction, hydrolysis or transfer of molecules.

Myrioid of reactions involved in the intermeditary metabolism of carbohydrate proteins and fats, intracellular oxidation and reduction reaction, CO_2 and oxygen transport, muscle contraction, blood clotting, excretion, sections calciflication or detoxyfication are controlled by various enzymes. The efficiency of an enzyme is expressed in terms of their turn over number *i.e.* number of substrate molecules acted upon by a single enzyme. The enzyme activity is further a function of specific activity temperature controlling to attain specific activation energy level of reacting molecules. Untill the reacting molecules altarn the level of activation energe these donot react. For example one molecule of uriase can breakdown 30,000 molecule of urea per second and a carbonix anhydrase molecule can act upon 36 million molecules of carbon dioxide to form carbonic and molecule with water per minute.

6.1 Digestive Enzymes

Food is largely desired from animals or plants or both. Much of our food is composed of large molecules such as "starch protein" and lipids which can not be absorbed from the intestine into the blood stream and could not be utilized if they were absorbed. In fact they might prove toxic if undigested. Consequently, various large molecules are broken into these constituents of monomers such as starch into glucose molecule using the proteolytic enzymes by the enzyme amylase, proteins into amino acid and lipids into fatty acids by lipids. These simple monomere are absorbed into the blood and distributed to various organs where they are again recombined into required macromolecule. In addition, the digestive system also provide entry to water, mineral salt, vitamins and many other differable molecule into the blood streams to be ultimately utilized inside tissue/cells. A few proteolytic enzyme are also reported to occur in plants/fruits like papaya the milky tune of which contains papain. Bromelin is present in pine apple juice and ficin in fig latex. Urease is reported to be present in dark beans and some other seals.

A variety of phosphatase exists in animal tissue such as phosphomaoesterases, phosphodiesterasis and phosphorylases. The acid and alkaline phosphatases of serum are so called because of the optimum activity at pasturlal pH. The determination of these enzymes is important in a number of clinical conditions such as phosphate gland phosphatase has been distinguished from other serum phosphal ares. Acetyl choline a mediator of nerve impulse is hydrolysed by the cholineesterase. A variety of other enzymes are also known for diagnostic purposes.

Action of Amylase

Starch molecules are the complex polymeres of glucose. On mixing with an amylase and the other major part remain insoluable an amylopactin. On the hydrolysis both these compounds are broken into glucose units. The amylose molecule is different from the amylopactin due to their unbranched chain of glucose molecules and relatively lower molecular weight. Prytopactin molecules are branched structures with much higher molecular weight (50,000 to 100000). The enzymes that hydrolysis starch are known as amylases and are present in animals as amylase (ptylin in saliva and pancreas juice) and as amylase (wheat and barley seed). Amylase first act upon amylose to split it mainly into small dextrin molecule with little maltose. Dextrin is further hydrolysed as maltose where as B amylase hydrolysis mylose completely into maltose. Both and B amylases act upon mytopacting in a way that these supplement each other action.

Amylase component of starch gives blue black colour. With iodine where as destrin and amylopactin molecules gives red violet colour with iodine glycogen also red brown colour with iodine.

Glucose is acted upon by β-amylase. Cellulose, other complex polysaccharide is digested only by the action of cellulase present in various bacteria and some other lower forms like fungi. Among animals only few forms like the snail *Helix pumation* produce cellulase in their digestive juices. The disaccharide sugars like sucrose, maltose and lactose do not exist in blood if these are injected directly in the blood, these appear excreted immediately in the urine, liver regulate blood sugar levels. If the liver is removed, an animal is unable to survive due to sudden fall in blood sugar levels (hypoglycemia) as the muscle glucogen can not be utilized as a direct source for glucose. Insulin stimulate both sugar penetration into cells and tissue and also glycogen production (excessive removal of sugar in urine). Likewise oxygen stress asphyxia also result in loss of liver glycogen. The steroids further enhance blood glucose levels due to glycones governs. The anterior pituitary secretes certain hormones resulting in hyperglycemia through adrenal cortex and the adrenal medulla through epinephrin also caused hyperglycemia under stress or emotional disturbances. These effects primarily are in both liver and muscle glycogen.

Exp. 43: Determination of the Optimum Temperature of ∝-amylase

Principle

∝-amylase catalyses the hydrolysis of 1-4 links of starch with the production of reducing sugars. The reaction is followed by measuring the increase in reducing sugars using the 3,5-dinitrosalicylate reagent when an alkaline solution of 3,5-dinitrosalicylic acid is reduced to 3-amino-5-nitrosalicylic acid. The reaction is followed by measuring the extinction at 540 nm.

Materials

1. Sodium or potassium phosphate buffer (0.1 mol/l, pH 6.7).
2. Buffered starch substrate (5g/l) in phosphate buffer. Mix 5 g of soluble starch to a smooth paste with about 50 ml of buffer solution. Add this quantitaively to 500 ml of boiling phosphate buffer solution, continue to boil for 1 min, then cool to room temperature, and dilute to 1l with buffer solution.

3. Sodium chloride (10 g/l)
4. ∝-Amylase suitably diluted.
5. Sodium hydroxide (2 mol/l).
6. Colorimeter
7. Water bath at a range of temperatures upto 75°C.
8. Sodium potassium tartrate (Dissolve 300 g of this salt in about 500 ml of water).
9. 3,5-dinitrosalicyclic acid (Dissolve 10 g of this reagent in 200 ml of 2 mol/l sodium hydroxide).
10. Dinitrosalicylate reagent (prepared this by mixing solutions 8 and 9 and making upto 1 l with water).

Method

1. Set up eight tubes containing starch (5g/l) 2.5 ml, Phosphate buffer (0.1 mol/l, pH 6.7) 1.0 ml and sodium chloride (10g/l) 0.5
2. Place the tubes in a water bath at a fixed temperature and equilibrate the enzyme at the same temperature.
3. After 10 min add 0.5 ml of the enzyme to seven of the tubes and 0.5 ml of water to the blank.
4. Incubate the tubes for 0,5,10,15,20,30 and 40 min and stop the reaction by adding 0.5 ml of 2 mol/l NAOH to all tubes.
5. Add 0.5 ml of the dinitrosalicylate reagent and heat the tubes for 5 min in a boiling water bath.
6. Cool, and read the extinction at 540 nm against a blank
7. Plot the extinction of each solution against the time of incubation and prepare a progress curve of the reaction.

Collect the data from the other groups and plot a graph of the change in the initial reaction rate against temperature and also the substrate changed per minute at 10 and 40 min. against temperature. Compare these graphs and explain them.

Plot a graph of log 10 v against 1/T and determine the heat of activation. What assumptions have been made in this experiment.

For convenience the activity can be expressed in terms of extinction change at 540 nm per minute.

Exp. 44: The Effect of Temperature on the Stability of ∝-amylase

Materials

The reagents and quantities are the same as in the previous experiment.

Method

1. Prepare a suitable dilution of the ∝-amylase so as to give a good activity when incubated 5 min at 37°C.
2. Place the enzyme in a boiling tube and equilibrate for 10 min.
3. Pipette 0.5 ml of the enzyme into four test tubes in a water bath at a fixed temperature from 50 to 75°C.
4. Remove the tubes after 5,10,15 and 20 min (or other suitable time intervals) and rapidly cool under running tap water.
5. The mixture is incubated for 5 min and the activity determined with 3,5-dinitrosalicylate. The rate of inactivation of the enzyme can then be determined.
6. Prepare a graph of the rate of activation (v) against 1/T and determine the heat of inactivation. Compare this figure with that obtained for the heat of activation from the last experiment.

Exp. 45: pH and Enzyme Activity

Enzymes are active over a limited pH range only and a plot of activity against pH usually gives a bell-shaped curve. The pH value of maximum activity is known as the optimum pH and this is a characteristic of the enzyme, provided that the enzyme is stable under the conditions studied.

The variation of activity with pH is due to the change in the state of ionization of the enzyme protein and other components of the reaction mixture is active so that a change in pH either side of the optimum of this form and hence a fall in the activity. If the enzyme is unstable at certain pH values, then the optimum pH is no longer a characteristic of the enzyme. The stability can be checked by exposing the enzyme to the appropriate pH value for the time of the experiment and then adjusting the pH to a value at which the enzyme is stable and measuring the activity.

Unit–III

Exercise No. 1

Introduction

In this unit concise descriptions of the techniques and methods of water analysis have been discussed. It contains exercises and comprises a comprehensive range of basic experiments, suitable for teaching the fundamental concepts of practical farming of fisheries. Although, many simple procedures have been omitted, but some basic procedures have been retained, which would be useful when analysis would be done with limited facilities at the field and laboratory by the students. It also contains informations on reagents and analytical steps have been set out with specific details, bearing in mind the need of the workers at bench. Therefore, in many cases, the analytical methods can be adopted to a wider range of concentrations by appropriate dilution of sample of water or titrant solutions or by varying the quantity of sample taken for analysis. In this manual, exercises elucidate the basic concept of various fishes and animal biochemical and physiological aspects.

1.1 General Precautions

In this unit a selection of analytical methods is described, dealing with the constitutions most often required in a chemical analysis of lake, pond or river water. The constituents analyzed may be irregularly distributed within the sampled water-body. Such sampling variance can especially be large for particulate material, which may include the larger zooplankton and also dislodged fragments and silt load in running waters. Any such sampling should be from homogeneous (*e.g.* well-shaken) water.

The sample may be contaminated by the water sampler itself or by the final recipient vessel. When sampling for nutrients which may increase steeply in concentration in the lower water layers, it is best to collect samples from the surface downwards. Care should be taken to avoid mechanical problems of a depth-sampler, such as miss-firing at the wrong depth, leakage, and disturbance of bottom.

Contamination may occur from glassware used during the analysis, or from reagents, including distilled water, or from filters. Reagent-contamination is often compensated by subtraction of 'blank' determinations with distilled water. Natural samples (sometimes turbid or coloured) plus reagents can also be used as 'blanks' if one critical stage or reagent is omitted.

Concentrations under analysis, in an unknown sample, may be incompatible with the reagent quantities and concentrations or the final read-out used. This can often be remedied, after a preliminary trial, by a known dilution. As an initial guide, measurements of electrical conductivity–closely related to total ionic content - are often useful in planning further analysis, especially of more saline waters. In some instances dilution is inapplicable and reagent concentration must be raised (*e.g.* ferrous iron determinations). In general, higher scale (>0.8 units) of absorbance in spectrophotometer determinations, especially with non-digital instruments, should be avoided.

1.2 Sampling

One of the simplest sampling devices is a bottle, weighted sufficiently to sink in water when empty. This is lowered through a synthetic thread or wire, tied firmly round its neck and also attached about 10 cm higher up to the stopper. A jerk on the line, when the bottle is at the required depth, removes the stopper. This is adequate for many purposes in shallow water, but will not provide samples for the determination of dissolved gases.

For water sampling in lakes, ponds or deep rivers, it is sometimes desirable to obtain an integrated sample representative of the uppermost 5 or 10 meters of water. This may be done by using a flexible plastic tube (polyvinylchloride) of about 2cm bore and appropriate length; the tube is weighted at one end and open at both the ends. The weighted end to which the cord is attached is lowered slowly into the water so that when the tube is fully extended, it encloses a columnar section of the upper water. The upper end of the tube is then closed by a stopper and the lower end is raised by the attached cord. The enclosed water is then run into a sample bottle and mixed by shaking. The large samples may be taken by means of a water pump. To avoid contamination, one can use a peristaltic motor or hand driven pump, or a simpler air lift pump.

1.3 Storage and Preservation of Samples

Even for short-term storage of water samples, soda glass bottles should be avoided, since a contact period of more than a few hours may significantly alter the concentration in the water sample of sodium, calcium and silicate. For most purposes the best materials for sample bottles are polyethylene or polypropylene. These plastics have advantageous properties both physically and chemically. Bottles made from them are virtually unbreakable and are light in weight. There is little danger that inorganic substances (excepting trace metals, especially zinc) will be extracted into water contained in them, but there is a risk with organic substances. An acid wash (*e.g.* soaking with 10 per cent nitric or hydrochloric acids for 48 h) is advisable for all new bottles, with periodic scrubbing after use.

Freezing the water sample in a plastic bottle is an effective method of preservation for most subsequent analyses apart from dissolved gases, pH, and soluble reactive silicon. However, adverse effects on determinations of soluble reactive phosphorus and alkalinity have also been reported, and concentrations of Ca^{2+} can be altered in calcium-rich waters. If necessary, freezing can be done at the collection site by immersing the bottle in a mixture of solid carbon dioxide and alcohol contained in a deep-freezers. Before analysis the samples should be thawed completely and mixed. For short (*e.g.* overnight) periods, unfrozen samples may be stored at 4°C.

Chemical preservations are often used, but may disqualify the sample for some analyses, and are poor substitutes for early analysis. There is no universally effective and suitable preservative. Mild acidification (*e.g.* to pH 2-3, with nitric or sulphuric acids) reduces the risk of precipitation or adsorption of metal and other cations (*e.g.* Fe^{3+}, Cu^{2+}, Ca^{2+}, Mg^{2+}) during storage, but some original distinctions (*e.g.* between ferrous and total iron) are lost, as are the alkalinity and conductivity, and nitrate or sulphate. The additive most used to repress biological activity is probably mercuric chloride, in concentrations of 40-70mg $HgCl_2\ l^{-1}$ (*e.g.* 1 ml of a near-saturated solution to 1 liter of sample). However, this interferes in determinations of ammonia, nitrate and, of course, chloride. It can be useful in arresting changes of total carbon dioxide. In some analyses, storage is possible at an intermediate stage, as after evaporation to dryness or after addition of one or more initial reagents.

1.4 Particulate and Soluble Material

A sample of natural water will contain significant quantities of many elements that occur both in solution and in solid (particulate) form. Analyses may be intended for one or both of these fractions, or their sum. The last usually requires the particulate component to be brought into solution; the processes include oxidative or acid digestion (iron, phosphorus, organic nitrogen, and organic carbon) and alkaline fusion (silicon). The analyses of some dissolved constituents (*e.g.* silicon, alkalinity components–at least in CA-poor, 'soft', waters) can usually be carried out with little risk of unwanted contributions from the particulate fraction, but for others (*e.g.* soluble reactive phosphorus) a preliminary filtration is necessary. Significant contamination from the filter may occur, especially for illiterates of waters with low ionic content. It must be assessed, by treatment of 'blank' filters, in measurements of elements (*e.g.* C, N and P) in particulate material retained on the filters.

The solution obtained in this way is suitable for determination of most metallic and some non-metallic components, but not for such elements as sulphur, halogens and possibly boron. For these latter elements, separate treatment of the original material must be made. Halogens and boron may, for example, be determined on solutions prepared by heating the material with sodium carbonate in borosilicate test tubes, or with sodium hydroxide in nickel vessels, extracting the resultant material with water, centrifuging, and then using the supernatant solution for analyses.

1.5 Glassware

For most analysis, borosilicate glassware should be used. When traces of metals or boron are sought, quartz vessels are desirable. Time spent in trace analysis using contaminated glassware is time wasted. It is therefore important to be certain that all vessels coming into contact with samples or reagent are clean. Glassware may be cleaned by immersion for several hours in the conventional chromic-sulphuric acid mixture, which is most effective but hazardous when hot. It is essential to wear goggles and rubber gloves when using corrosive cleaning mixtures. More convenient and innocuous surfactant cleaning agents are now commercially available (*e.g.* Decon 90, RB 75) and are adequate for many purposes; a final acid wash (*e.g.* with dilute HCI) is often helpful.

After use of cleaning mixtures, the glassware should be well rinsed in distilled or deionized water. Tap water should be used with caution particularly when trace metal analysis is contemplated, since most tap waters are heavily contaminated with, for example, copper and zinc derived for the piping. Do not touch the inner surfaces of cleaned glass water, as fingers are usually richly contaminated with many elements, particularly heavy metal derived from brass taps and the fittings, and sodium and chlorine from perspiration.

1.6 Reagents

Use analytical grade chemicals (AR) for the preparation of reagents. Always run a "blank sample" on the spectrophotometer, in this the sample is usually replaced by distilled water.

1.7 Colorimetry

A number of analyses depend upon the quantitative measurement of the density of colour in a solution, by means of a spectrophotometer. If the latter is not available, simple filter-colorimeters have an advantage of reproducibility. The older technique of visual colour match requires still less equipment, and may be useful in field situation *e.g.* where larger errors are likely to be introduced by a delay between sampling and analysis. The process of visual colour matching of liquid columns, illuminated from below, is applicable to numerous colorimetric methods employed in water analysis. The principle is the same whether applied, as it is here, to comparison of sample and standard in simple Nessler tubes, or in more complexes colorimetric. When a colour match is achieved between treated sample and standard solution, the respective concentrations are inversely proportional to the column heights provided that the latter are not too dissimilar. Height may be assessed directly or from the volume poured out of the more deeply coloured solution. The apparent colour of artificial standard tests or tends vary with the source of light. A constant source of light, preferable strong diffuse day length reflected from a white card is advised. For field situations, forms of multiple colour glass or plastic standards carried on a rotating disc are conveniently compact and durable (*e.g.* Tintometer assembly). A "blank sample" in a Nessler tube is viewed through the colour disc, whose filter intensity is adjusted (By rotation-substitution) until it matches that of the unknown treated sample held parallel to the blank. Recently there has been increasing use of small portable photo electric colorimeters.

1.8 Ion-selective Electrodes

The concentration of hydrogen ions can be measured with the glass electrode, so that number of other ions can be measured by other ion-selective ('specific ion') electrodes. These are obtainable from the major firms manufacturing pH electrodes and meters, whose current brochures should be consulted. Electrodes for the following ions are of particular value to the water analyst: F^-, S^{2-}, Na^+, K^+, monovalent cations, Ca^{2+}, and NH^+_4, electrodes are available for many of the important ions, including NO^-, but do not necessarily have sufficient sensitivity and selectivity to cope with the very low concentrations that are common place in many natural water.

The result of many analyses has been expressed in terms of several chemical entities, which are sometimes quite unrealistic. Weight based concentrations related to such radicals of minor constituents a phosphate, nitrate, and ammonium are best expressed in terms of the main element concerned. To avoid confusion, this should be shown explicitly (*e.g.* PO_4-P, NO_2-N, NH_4-N). By a chemically unrealistic convention dissolved reactive silicon is often expressed as concentrations of its oxide, silica (SiO_2) which is mainly significant as a product of utilization. Concentrations involving Ca^2 and Mg^2+ and their sum, have been expressed in various units of hardness.

1.9 Safety

Many chemicals and instruments used in a laboratory can provide a serious health risk to the people who work with them. Therefore, it is a common responsibility of everyone who works in the laboratory to protect him, and his fellow workers from accidents.

☆ Fire extinguisher should be placed near the entrance of the laboratory. Everyone working in the laboratory has to know how to use this equipment.

☆ Laboratory fume hood protects the workers in the laboratory from inhalation dangerous fumes. Use the fume hood every time you work with strong caustics or acids. Clean all spills immediately.

☆ Always wear a laboratory coat to protect you from contact with chemicals, and to protect your normal clothes from being damaged.

☆ When working with strong caustics or acids always wear rubber gloves.

☆ When there is mouth-pipetting, there is always a risk of swallowing a small amount of reagent. To prevent this, always use a pipetting balloon when the reagent can be dangerous to your health.

☆ Despite all precautions a reagent contacts the skin or eyes, immediately flush with excessive water. If irritation persists, seek medical attention immediately.

Cleanliness

(*a*) Environment

☆ Keeps your laboratory working area clean and tidy. A clean laboratory not only looks better, but you also protect your glassware, equipments and other materials from being contaminated with dust.

☆ If you spill reagents when working with it, clean all spills immediately, as many chemicals are corrosive, and thus can damage tables, or worse expensive equipment and also many chemicals are dangerous to your health.

(*b*) Glassware

☆ To obtain the exact data, it is essential that the glassware to be used should not contain even a fraction of the dust or the chemicals. Therefore, modern techniques must be employed for cleaning of glassware.

☆ Glassware should be cleaned immediately after use. Soap and scrubbing with a brush is generally not necessary.

Rinsing with IN HCL is good for all glassware but especially for glassware used to determine ammonia and phosphorous because it removes N and P quickly.

☆ The following procedure is proposed for cleaning glassware.

(*i*) Rinse three times with tap water.

(*ii*) Rinse 1 time with distilled water.

(*iii*) Rinse 2 times with 1 N HCl.

(*iv*) Rinse 2 times with distilled water.

(*v*) Dry in an oven and store upside down or cap with aluminium foil to prevent contamination with dust particles.

(*vi*) Finally Autoclave all the glassware before use in the laboratory.

Pipetting

☆ When exact volumes have to be measured *i.e.* in making standard solutions, one has to make use of a pipette. The most accurate one is the volumetric pipette which will be used in most cases.

☆ Always follow the procedures below to obtain the highest precision.

Volumetric Pipette

☆ Rinse the pipette 2 times with distilled water (for phosphate analyse, first rinse the tip with 1 N HCl).

☆ Pour a small amount of the liquid to be pipetted into a clean and dry breaker.

☆ Suck a small amount of the liquid into the pipette and rinse with this liquid by holding the pipette horizontally and turning it. Discard the liquid from the pipette into the sink and repeat another two times.

☆ Fill the beaker with fresh liquid and such liquid into the pipette until a few centimeters above the ring mark. Close the pipette with the second finger.

☆ Wipe the tip dry with a tissue hold the pipette vertically with the tip in a 45 degree angle to the wall of the beaker and slowly let the liquid flow out of the pipette by lowering the tense on the second finger. Stop the flow when the underside of the meniscus corresponds with the ring mark (to do this correctly the eye has to be on the same height as the ring mark).

☆ Clean the pipette after use.

Titration

☆ Preparing the burette before use.

☆ Open the tap of the burette and rinse the inside of the burettes 3 times with distilled water.

☆ Place a clean and dry funnel on top and rinse the burette 3 times with ca. 10 ml. of titrant.

☆ Close the tap and fill the burette.

☆ Adjust the titrant level to somewhere in the beginning of the scale.

Titration Procedure

☆ Prepare the solution to be titrated according to the manual.

☆ If available, place the flask or beaker on a magnetic stirrer under the burette. If not available, it has to be swerved well during titration.

☆ Read the starting level of the titrant, keeping your eye at the same height as the meniscus. Some burettes have a so called Schellbach-stripe, a blue line behind the scale. This line forms a thin and a thick arrow at the meniscus with the points against each other. Read the scale where the points touch each other.

☆ Start adding titrant drop wise until first partially colour change occurs.

5. Add the last drops very slowly until complete colour change occurs which is stable for about 1 minute. When you are not sure you reached the end point yet, read the titrant level before adding one extra drop.

☆ Read the end level of the titrant and also notate the exact normality of the titrant as given on the reagent bottle.

☆ After finishing all titrations drain the burette completely and discard this titrant ever pours it back into the reagent bottle.

☆ Rinse the burette and store it completely filled with distilled water.

Reagents

☆ Always follow directions carefully. Remember to guard against the contamination of dry chemicals. If the bottle says that the chemical is "anhydrous" keep the cap on! Moisture from the air will contaminate it. NaOH is one chemical to be particularly careful with, as moisture absorbs rapidly.

☆ Use extreme caution when handling strong acids, caustics and phenol. Always pour acid into water, never water into acid. When chemicals arrive at the laboratory write on the label the date of arrival, and the date, the bottle was opened for the first time.

☆ For newly prepared reagents always make a label including:

(*a*) Name of reagent

(*b*) Date it was prepared

(*c*) Who prepared it?

☆ Some reagents need to be kept cold and/or in the dark. Therefore, don't expose them longer than necessary to harmful conditions. Never put a pipette into a reagent bottle. That is a good way to contaminate the reagent. Discard the liquid that is left in the beaker or flask. Never return it to the reagent bottle.

Colorimetric Determinations

☆ In the determination of water quality parameters many methods make use of colorimetric comparison between samples and standards.

☆ The use of a spectrophotometer offers a very precise way to do such comparisons, but only if this is done in the right way.

Making a Standard Curve

- ☆ Measuring the concentration of a certain parameter in a water sample is done by comparing its absorbance, after adding colour developing reagents, with the absorbance of a series of known standard solutions.
- ☆ In most cases a series of standard solutions is made for each time a series of sample concentrations has to be determined to obtain the highest precision. Only in the case of total phosphate determination this is time consuming, and a new standard curve is made only when one of the reagents is renewed.
- ☆ Switch on the spectrophotometer to warm up, and set the correct wavelength.
- ☆ Prepare all standards very precisely according to the determination manual.
- ☆ Add colour developing reagents and wait for the time given.
- ☆ Set full scale (100 per cent absorbance) of the spectrophotometer without a cuvette in the instrument.
- ☆ Set zero absorbance with a clean cuvette filled with distilled water in the instrument.
- ☆ Rinse the cuvette 3 times with a small amount of the 0.0 ppm standard, fill it with the standard, wipe dry and clean with a soft tissue and place it in the instrument. Read and note the absorbance.
- ☆ Repeat for the other standards. Check the zero adjustment in between each standard.
- ☆ Plot all absorbances readings against the standard concentrations on graphic paper and check if a straight line can be drawn through these points.
- ☆ Calculate a regression line for your standards.

Evaluating the standard curve

- ☆ A correct standard curve is straight and goes through the origin.
- ☆ It is very difficult to draw a straight line because the points are too scattered. This mostly occurs when the glassware

used was contaminated or the cuvettes used were not neatly clean. Clean all glassware thoroughly, check the cuvettes for fingerprints or other dirt and make a new standard curve.

- ☆ The curve is linear but does not go through the origin. The distilled water used in making the standard, or one or more of the reagents may be contaminated.
- ☆ If the contamination is due to the reagent, the standard curve can be used without correction because the contamination is assumed to be equal in both the standards and the samples. To test the reagents, run sets of standards doubling the volume of each reagent one at a time. If a reagent is contaminated the absorbencies should increase when the volume of that reagent is doubled. If the absorbencies do not increase with increasing reagent volumes, then it is most likely the distilled water is contaminated.

Exercise No. 2

Estimation of Physical Characteristics of Water Body

2.1 Estimation of Surface Temperature of Water

Temperature is one of the most important factor and all life processes are accelerated or slowed down by temperature changes in the environment. It influences the solubility of gases and slats in water. The volume as well as density of water depends upon temperature. The instruments which are used to measure temperature are mainly modifications of mercury expansion thermometers. In recent days, many electrical devices like thermocouple and thermister are used.

Measurement of Water Temperature

Measurement of temperature at various depths below water surface requires specially designed instruments for this purpose. Temperature often differs with depth and positions at which temperature is to be measured are remote from the observer. So it is necessary to understand the possible errors, such instruments when lowered, meet rapidly increasing hydrostatic pressure and so may require correction for pressure as well as temperature.

2.2 Reversing Thermometer for Recording Temperature at Different Depths of Water

This instrument consists of the main thermometer which registers the temperature and an ordinary thermometer which is used for correcting the change brought about by atmospheric temperature. The main thermometer has a mercury reservoir at one end, which is connected to a graduated capillary tube by a capillary loop with a construction. At the other end of the tube, is an empty reservoir. For use, the thermometer frame is lowered into the water vertically to the desired depth, after a few minutes, a message is sent down the rope which reverses the thermometer frame. This reversing action breaks the mercury column at the point of construction and the mercury fills the empty bulb and a portion of the capillary tube. The level of mercury in the capillary indicates the temperature of water at that depth.

2.3 Thermister for Continuous Recording of Water Temperature

This consists of a small capsule containing a complex metal oxide compound whose resistance decreases rapidly with increasing temperature. When used in a simple whetsotne bridge circuit, it acts as a sensitive thermometer, and has the great advantage of being continuously recorded. For each measurement, the valuable resistance R_3 is adjusted until no reading is obtained on the galvanometer. When finding the temperature of water, the Thermister (T) can be attached to a calibrated rod and low resistance leads fitted to it. It is important that these and all joints are well insulated with water proof material. The valuable resistance R_3 is provided by a rheostat fitted with a degree scale, while a small torch battery (B) and relatively inexpensive moving coil galvanometer (G) are the additional requirements.

2.4 Calibration

A simple conventional graph must be constructed in the laboratory by immersing the Thermister in water of varying temperatures (measured by sensitive mercury thermometer) and plotting against corresponding values of R_3. It is important that while doing so, same leads from the Thermister should be used as in the field.

2.5 Estimation of Turbidity of Water

In any water body, some materials occur in undissolved conditions in the later and due to presence of these undissolved materials, water becomes turbid. The turbidity of water may more or less; depend on undissolved materials which are present in the water.

If these materials are more, the water will be more turbid. The turbidity of water is directly related to light penetration and visibility. The turbidity of water can be measured by Secchi disc which is a simple device used to measures turbidity and visibility of water. It consists of a weighted circular plate, 20 cms in diameter, with the surface painted with opposing black and white quarters. It is attached to a calibrated line by a ring at the centre so that when held by line, it hands horizontally.

Procedure

To determine the water turbidity, depth of 'Secchi Disc' visibility is determined. To determine the above, slowly lower the disc into the water until it disappears, and notes this depth (A). Lower the disc a few more feet, and then slowly raise it, until it reappears, and note this depth (B). The average of these two readings is taken for the final Secchi Disc visibility depth, *i.e.* (A+B/) = visibility depth.

2.6 Estimation of Specific Conductivity of Water

Specific conductance is a measure of water's capacity to conduct an electric current. It is the reciprocal of resistance (R) for which the standard unit is an ohm. Since conductance is the inverse of resistance, the unit of conductance is mho, or in low conductivity natural waters, the measurement is taken in micro mho. Because the measurement is made using two electrodes placed 1 cm apart, specific conductance is generally reported as micro mho per cm. The relationship of specific conductance to ionized matter concentration varies with both the quality and quantity of the ions present. However, at low concentrations, the ions move and behave independently, and the conductance–concentration relationship is almost linear. Temperature of the solution affects the ionic velocity, and thus, the specific conductance.

The greatest value of specific conductance to the limnologist is the estimation, by this very simple procedure, of the total concentration of dissolved ionic matter in the water, which in turn, is related to water fertility. Total dissolved matter (TDM) may be estimated from conductance data, by multiplying some empirical factor, which usually varies from 0.5 to 1.0 (This method is considered under the procedure for dissolved salts).

Apparatus Required

- ☆ Conductivity meter with dip cell.
- ☆ Water bath at 10°C or 25°C, if conductivity meter lacks temperature compensation.
- ☆ Thermometer.
- ☆ Two beakers of 200 to 250 ml capacity.

Procedure

Connect plugs of dip cell to instrument (the plugs may be connected to either post). A cell constant of 1 is best for most natural waters' other cell or ranges are available for weak or strong conductors. Fill two breakers with water sample to a depth, sufficient to cover the vent holes in the dip cell when immersed. Place in water bath and wait for equilibrium, if necessary. Measure temperature of the second beaker and set temperature compensator on instrument to that temperature. Rinse dip cell thoroughly in the first beaker, and then transfer dip cell to second beaker. Turn on the instrument, and while gently move the dip cell, keeping the holes submerged all the times, take a reading of specific conductance. Turn off the instrument, remove and disconnect dip cell, rinse with distilled water and air dry.

Exercise No. 3

Estimation of Chemical Characteristics of Water Body

3.1 Estimation of pH of Water by Different Methods

The pH (Hydrogen ion concentration) of a solution is a measure of its hydrogen ion activity and is the logarithm of the reciprocal of the hydrogen ion concentration. Thus, it is important to remember that a change of one pH unit, represents a tenfold change in hydrogen ion concentration, *i.e.* a pH of 6 has 10 times the hydrogen ions of pH 5 has 100 times the hydrogen ion of pH 7.0. The pH of most natural waters falls in the range of 4 to 9 and much more often in the range of 6 to 8. The following methods are used for measurement of hydrogen ion concentration:

- ☆ Colour chart method
- ☆ Universal indicator solution method
- ☆ pH meter method

3.1a Colour Chart Method

In this method, sets of graded colours representing range of pH values are used. Such colours usually printed on paper in some convenient form.

Procedure

- ☆ Take a pH indicator paper strip and dip it into given water sample and compare a colour of strip with given graded colours chart. This gives approximate pH values.

3.1b Universal Indicator Solution Method

In this method, universal indicator solution is used. A few drops of this solution are added into the sample. This develops a colour in sample. The colour of sample is compared with given standard chart. Indicator solution develops different types of colours at different pH in sample. This gives an approximate value of pH.

3.1c pH Meter Method

Apparatus

- ☆ pH meter
- ☆ Centigrade thermometer
- ☆ Beakers
- ☆ Blotting paper

Reagents

- ☆ pH buffer solution in the range of water sample being analyzed.
- ☆ Distilled water

Procedure

Prepare a buffer solution with a pH near the expected pH of the water sample (buffer may be prepared in the laboratory or may be purchased from any firm). Place 25 ml buffer solution in a small beaker. Open the vent of glass reservoir of the fiber junction electrode. Rinse the electrodes with distilled of the fiber junction electrode. Immerse the electrodes in the buffer solution, measure temperature of buffer solution, and set compensator to that temperature. Set function switch to pH, and adjust meter needle to the pH of the buffer solution. The instrument is now standard. Return switch to standby. Rinse electrodes in distilled water and blot dry. Immerse electrodes in sample, measure temperature of sample, and set compensator to this temperature. Set function switch to pH, and very gently swirl (or use magnetic stirrer at very slow speed) until pH meter needle stops drifting. Read pH from meter scale. Return switch to standby. Raise electrodes, rinse with stilled water and dry. The instrument is now ready for the next sample or for storage. Storage of instrument–Close the glass reservoir to minimize electrolyte loss, and leave electrodes immersed in distilled water.

3.2 Estimation of Alkalinity of Water (Carbonates and Bicarbonates)

The alkalinity of water is its ability to accept protons; stated another way, it is the quantity and kinds of compounds present, that collectively shift the pH to the alkaline side of neutrality. Although the alkalinity of natural waters is generally the result of bicarbonates, it is usually expressed in terms of calcium carbonate. Three kinds of alkalinity are indicated: hydroxide (OH), normal carbonate (CO_3^-) and bicarbonate (HCO_3^-), the three is summed as total alkalinity. Carbonates and bicarbonates are common to most waters because carbonates, minerals are abundant in nature and the contribution to alkalinity by hydroxides is rare in nature. The presence of hydroxides can usually be attributed to water treatment or to contamination. Expected total alkalinities in nature usually range from 45 to 200 mg/l.

Principle

There are five conditions of alkalinity possible in the sample: carbonate, bicarbonate, or hydroxide alone, or a combination of carbonate and hydroxide or carbonate and bicarbonate. Hydroxide and bicarbonate are not found together. Alkalinity is determined by titrating the water sample with a standard solution of strong acid. The equivalency or end points of the titration are selected as the inflection points in the titration of sodium carbonate with H_2SO_4. These points will vary slightly with temperature, ionic concentration, and free carbon dioxide concentration. This end point may be determined empirically by titration and is the pH where the derivative of pH/ml titrant is greatest. However, they are usually taken as pH 8.3 for the carbonate end point and pH 4.5 for the bicarbonate end point. The following reaction occurs in titrations:

$CO_3 = + H^+ \longrightarrow HCO_3^-$ (Titration to pH 8.3)

$HCO_3^- =$ (from $CO_3 =$) $+ H^+ \longrightarrow H_2O + CO_3$

HCO_3^-(Natural) $+ H^+ \longrightarrow (H_2O + CO_2)$ (titration to pH 4.5)

Apparatus Required

Phenolphthalein Indicator Solution

Dissolve 0.5 g phenolphthalein in 50 ml of ethyl alcohol and add 50 ml of distilled water. Then add 0.02 N NaOH drop wise until a faint pink colour appears.

Standard Sulphuric Acid Solution

Prepare stock solution, approximately 0.1 N by diluting 3ml of concentrated sulphuric acid to 1 liter. Dilute 200 ml of 0.1 N stock solutions to 1,000 ml with carbon dioxide free water to give 0.02 N solutions. For accurate work, this must be standardized against Na_2CO_3.

Methyl Orange Indicator Solution

Dissolve 0.5 g methyl orange to 1 liter distilled water.

Procedure

Phenolphthalein Alkalinity

Take 50 or 100 ml of water sample and add 4 drops of phenolphthalein indicator solution. If a pink colour appears, (indicates the presence of carbonates) titrate it against standard sulphuric acid (0.02N) until colour disappears (pH 8.3) and record milliliters of acid used during the titration.

Total Alkalinity: (by Methyl Orange Indicator Method)

If no pink colour appears after adding phenolphthalein or after phenolphthalein titration, add 2 drops of methyl orange indicator and titrate to a faint orange at pH 4.6 (and pink at below 4 pH). Record total used standard acid in ml.

Calculation

$$\text{Phenolphthalein Alkalinity as mg } CaCO_3 \text{ per liter} = \frac{\text{Ml Standard Acid}}{\text{ml of Sample}} \times 1000$$

$$\text{Total Alkalinity as mg } CaCO_3 \text{ per liter} = \frac{\text{Ml Standard Acid}}{\text{ml of Sample}} \times 1000$$

3.3 Estimation of Free Carbon Dioxide of Water

Carbon dioxide is an end product of decomposing bacteria and of the respiratory process of plants and animals. It is also added to the water by the action of natural or pollution acids on bicarbonates.

Principle

Free carbon dioxide reacts with sodium carbonate or sodium

hydroxide to form sodium bicarbonate. Completion of the reaction can be indicated by the development of the pink colour, characteristic of phenolphthalein indicator at the equivalent pH of 8.3.

Apparatus Required

- ☆ Burette
- ☆ Conical flask, etc.

Reagents Required

- ☆ Phenolphthalein indicator solution
- ☆ Preparation of phenolphthalein indicator solution is same as in estimation of alkalinity.
- ☆ Standard sodium carbonate titrant (0.045 N)

Dissolve 0.602 g anhydrous, primary standard grade Na_2CO_3 that has been oven dried at 140°C overnight, in freshly boiled and cooled carbon dioxide free water, and dilute to 250ml, store in a rubber stoppered Pyrex bottle (This solution should not be kept more than 1 to 2 weeks).

Procedure

Take 100 ml sample in Nessler tube and add 10 drops of phenolphthalein indicator. If the colour of the sample turns to pink, free carbon dioxide is absent. If the sample remains colourless, titrate rapidly with standard sodium carbonate titrant, stirring gently with a glass rod until a faint pink colour appears. Record the used volume of titrant in ml.

Calculations

$$MgCO_3 \text{ per liter} = \frac{\text{Used Vol. of Titrant}}{\text{ml of Sample}} \times 1000$$

3.4 Estimation of Dissolved Oxygen of Water (Winkler's Method)

The dissolved oxygen (DO) content of waters results from (1) the photosynthetic and respiratory activities of the biota in the open water, the benthos, and the autotrophs and (2) the diffusion gradient at the air interface and distribution by wind driven mixing.

Principle

This method depends on the formations of a loose precipitate (flock) of manganous hydroxide. The oxygen dissolved in the water is rapidly absorbed by the manganous hydroxide, forming a higher oxide, which may be in the following form.

$$MnSO_4 + 2\,KOH = Mn\,(OH)_2 + K_2SO_4$$

$$2Mn\,(OH)_2 + O_2 = 2\,MnO_2(OH)_2$$

As it settles, the Mn $(OH)_2$ flock acts as a gathering agent for the Dissolve Oxygen (DO). Upon acidification in the presence of iodide, iodine is released in a quantity, equivalent to the DO present.

$$MnO\,(OH)_2 + 2\,KI + H_2O = Mn\,(OH)_2 + I_2 + 2\,KOH$$

The liberated iodine is then titrated with a standard sodium thiosulphate solution, with starch used as the indicator.

Apparatus Required

- ☆ Narrow mouth dark or brown bottles of good quality with well fitting ground glass stoppers.
- ☆ Pipette 1 ml, 10 ml and 25 ml.
- ☆ Burette (50 to 100 ml capacity)
- ☆ Conical flasks etc.

Reagents

Manganese Sulphate Solution (Winkler–A)

Dissolve 120g $MnSO_4.4H_2O$ or 100g $MnSO_4.2H_2O$ or 91g $MnSO_4.H_2O$ in distilled water, filter and dilute to 250 ml.

Alkaline–Iodine (Winkler–B)

Dissolve 125g sodium hydroxide (NaOH) and 33.75g sodium iodide (NaI) in distilled water and dilute to 250 ml or dissolve 175g KOH and 37.5 KI in distilled water and dilute to 250 ml.

Starch Solution

Prepare an emulsion of 6 g starch in a beaker with a small amount of distilled water. Pour this emulsion in 1 liter of distilled water, boil for a few minutes, and let settle overnight. Save the clear supernatant, and preserve by adding 1.25 g salicylic acid.

Sulphuric Acid

Concentrated sulphuric acid is used.

Sodium Thiosulphate Stock Solution (1.0N)

Dissolve 248.2 g $Na_2S_2O_3.5H_2O$ in boiled and cooled distilled water and dilute to 1000 ml. Preserve by adding 5 ml chloroform or 1 g NaOH per liter.

Standard Sodium Thiosulphate Solution (0.25N)

Prepare by (a) dilute 25 ml sodium thiosulphate stock solution (1.0N) to 1000 ml or (b) dissolve 6.305g $Na_2S_2O_3.5H_2O$ in freshly boiled and cooled distilled water and dilute to 1 liter. Standard sodium thiosulphate solution may be preserved by adding 5 ml chloroform or 0.4 g NaOH per liter.

Standardization of Sodium Thiosulphate Solution with Bi-iodate

Dissolve approximately 2 g potassium iodide, free from iodate, in a flask with 100 to 150 ml distilled water; and 10 ml H_2SO_4 followed by exactly 20ml standard 0.025N bi-iodate solution. Dilute to 200 ml, and titrate the liberated iodine with thiosulphate solution, adding starch towards the end of the titration, when a pale colour is reached. Exactly 20 ml 0.025 N thiosulphate should be required when the solutions under comparison are of equal strength. It is convenient to adjust the solution to exactly 0.025 N. One ml 0.025 N thiosulphate is equivalent to 0.2 mg oxygen.

Procedure

- ☆ Fixation of dissolved oxygen
- ☆ Titration

Fixation of Dissolved Oxygen

- ☆ Fixation of dissolved oxygen must be performed in the field immediately after the sample has been secured. Any delay, results in error. Dissolved oxygen is fixed by following procedure.
- ☆ Take water sample in 250 ml or 300 ml volume sample bottle.
- ☆ Remove stopper from sample bottle and add 1 ml manganeous sulphate solution by inserting tip of pipette just below water surface.

- ☆ Add 1 ml alkaline–iodine solution in same manner as for $MnSO_4$.
- ☆ Replace stopper and mix sample by inverting bottle several times. Allow precipitate to settle for a few minutes. After settling of precipitate, a clear fluid should occupy upper portion of bottle.
- ☆ Using volumetric pipette, add 1 ml concentrated sulphuric acid by permitting it to run down neck of bottle; mix well by inverting bottle several times (at least 40 times). Allow sample to stand for at least 5-10 minutes.

Titration

Take 100 ml or 50 ml of sample (in which dissolved oxygen has been fixed) in a flask and titrate with 0.0025N sodium thiosulphate solution to a very pale yellow colour, now add 1 ml starch solution (colour of sample turns blue) and continue titration until the first disappearance of the blue colour. Record the used volume of sodium thiosulphate solution in ml.

Calculations

$$\text{mg of } O_2 \text{ per liter} = \frac{\text{Used Vol. of Titrant} \times 1000 \times 0.2}{\text{Vol. of Sample}}$$

where,

0.2 value represents 1 ml of sodium thiosulphate equivalent to 0.2 mg of O_2.

3.5 Estimation of Biochemical Oxygen Demand (BOD) of Water

The rate of removal of oxygen by organisms, using the organic matter, present in water, is called Biochemical Oxygen Demand (BOD). The parameter can easily be estimated, although only an approximation, but gives an index of organic pollution (or of sewage treatment plant efficiency). The BOD test is commonly made by measuring O_2 concentration in samples before and after incubation in dark at 20°C for 5 days. It is necessary to dilute and aerate the sample to ensure that not all the O_2 is used during the incubation. Sometimes a culture of bacteria is also added so that more of the organic matter is used up during incubation.

Method

A simplified method for BOD estimation gives some information about the difference between lakes or rivers or about season's changes in the dissolved organic matter in one water body.

Apparatus Required

- ☆ Narrow mouth, dark or brown glass bottles (125, 250, 300 ml).
- ☆ Air incubator or water bath thermo metrically controlled at 20 ± 1°C. All light should be excluded to prevent DO by algae in the sample.
- ☆ Burette, Pipettes, etc.

Procedure

If the O_2 content of one original sample is very low, the sample should be aerated for 5 or 10 minutes. Place portion of the sample in 3 glass stoppered bottles, the remaining bottle in the dark, at a standard temperature (*e.g.* 20°C) or at the temperature of the original sample for 1-5 days. To determine the remaining oxygen subtract this value from the original value of oxygen to give BOD

3.6 Estimation of Chemical Oxygen Demand (COD) of Water

Principle

The concentration of organic compounds in pond or lake waters can be estimated by their oxidability, by oxidizing substances such as potassium chromate ($K_2Cr_2O_7$). Organic carbon add O_2 are indirectly related, because the reaction is:

$$C_2H_2O_2 + x + \frac{(y - z)}{2} = O_2 - x\ Co_2 + y\,H_2O$$

The amount of carbon oxidized, can therefore, be calculated from the amount of O_2 taken up only if the ratio of (y-z) to O_2 is known. This ratio is, in most cases unknown. The COD method is, however, simpler than a carbon determination, and is directly comparable to the Biochemical Oxygen Demand (BOD), which estimates the organic compounds, available for bacterial activities. COD also has particular significance when photosynthesis is expressed in the same unit; as mg/I O_2. $K_2Cr_2O_7$ is the most suitable oxidizing agent.

The reaction is: $Cr_2O_7 + 14\ H^- + 6e^- == 2\ C1^{+++} + 7H_2O$ d

Ag_2SO_4 is used as a catalyst for the oxidation. It is particularly effective for short straight–chain alcohol and acids.

Reagents Required

- ☆ Standard oxalic acid, 0.0125 N
- ☆ $K_2Cr_2O_7$, 1.000N
- ☆ H_2SO_4 (*e.g.* = 1.84, A.R.)
- ☆ H_2SO_4 (1+1) A.R.
- ☆ Ferrous solution 0.25 N
- ☆ NH_2SO_3H (Sulphamic acid), (dry, A.R.)
- ☆ Diphenylamine indicator, 0.5 per cent
- ☆ Ag_2SO_4 (dry, A.R.)
- ☆ Hg_2SO_4 (dry, A.R)

Procedure

Mix 70 ml of sample containing not more than 16 mg COD (as O_2) or an aliquot diluted to 70 ml, in a 250 ml flask with powered Hg_2SO_4 (*e.g.* = 1.84, A.R.). Then add successively 100 mg of Ag_2SO_4 and 10.00 ml of $K_2Cr_2O_7$. Attach the condenser to the Erlenmeyer flask, and put the flask in a water bath. Either boils for 3-6 hrs. on autoclave the covered flasks at about 135°C (3 atmospheres pressure) for 3-6 hrs. Cool the flaks, and then, either titrate the excess of $K_2Cr_2O_7$ with Ferrous solution 0.25N. At the end point, the colour changes sharply from turbid blue to brilliant green. Run a blank using 70 ml of double –distilled H_2O and reagents.

Calculation

$$\text{COD mg/liter} = \frac{(a - b) \times N \times B \times 1000}{\text{ml of Sample}}$$

where,

COD = Chemical oxygen demand in mg/liter O_2

a = ml titrant used for blank (about 40 ml)

b = ml titrant used for sample

N = Normality of the titrant

3.7 Estimation of Chloride Contents of Water

Chloride is one of the major anions in water and sewage waters. The salty taste in water is produced due to the chloride concentration, which is highly valuable and depends on the chemical composition of water. High chloride content also exerts significant effect on metal-ice pipes and structures, as well as on agricultural plants.

Principle

In a natural or slightly alkaline solution, potassium chromate can indicate the end point of the silver nitrate titrant of chloride. Silver nitrate is quantitatively precipitated because, silver chromate is formed.

Apparatus Required

- ☆ Burette
- ☆ Conical flask etc.

Reagents Required

Silver nitrate solutions (0.0141 N): Dissolve 2.396g $AgNO_3$ in 1 liter distilled water.

Potassium chromate indicator: Dissolve 50g K_2CrO_4 in distilled water. Add $AgNO_3$ to produce a slight red precipitation. After it has stood overnight, filter and dilute to 1 liter with distilled water.

Procedure

Take 50 or 100 ml of water sample and add 4-5 drops of potassium chromate indicator solution. Now titrate the sample against the 0.0141 N $AgNO_3$ solutions till pinkish-yellow colour appears. Record the volume of used titrant in ml.

Calculation

$$\text{mg of Chloride per liter (mg/l)} = \frac{\text{Used Vol. of Titrant} \times a \times b \times c}{\text{Vol. of Water Sample}}$$

where,

a = 35.46 (atomic wt. of chloride)

b = 0.0141 (N of $AgNO_3$)

c = 1000 ml.

3.8 Estimation of Phosphate Contents of Water (Stannous Chloride Method)

Phosphorus in waters is present in several soluble and particulate forms, including organically bound phosphorus, inorganic polyphosphates and inorganic orthophosphates. All inorganic phosphates are usually considered as PO_4. Phosphorus is a biologically active element. Natural source of phosphorus to waters is from leaching of phosphate bearing rocks and from organic matter decomposition, additional sources are found in man-made fertilizers, domestic sewage and detergents.

Apparatus Required

- ☆ Spectrophotometer
- ☆ Acid-washed glassware (flasks & pipettes)

Reagents Required

Phenolphthalein Indicator Solution

(For preparation see section on estimation of alkalinity).

Standard Sulphuric Acid (0.02 N)

(For preparation see section on estimation of alkalinity).

Ammonium Molybdate Solution (Strong Acid Solution)

Dissolve 25.0g $(NH_4)6\ MO_{70}24.4H_2O$ in 175 ml distilled water. Continuously add 310 ml conc. H_2SO_4 to 400 ml distilled water, cool and add the molybdate solution and dilute to 1 liter.

Stannous Chloride Solution

Dissolve 2.5g of a fresh supply of $SnCl_2.H_2O$ in 10ml conc. HCl; then dilute to 100 ml with distilled water. Filter, if turbid (Store in cool).

Stock Phosphate Solution

Dissolve 0.7164g of potassium dihydrogen phosphate (KH_2PO_4)

(which has been dried in an oven at 105°C for 1 hour) in distilled water. Dilute the solution to 1 liter, 2ml of stock solution contains 0.5 mg of PO_4.

Standard Phosphate Solution

Dilute 100ml of stock solution to 1 liter, with distilled water. 1ml of this standard solution contains 0.05 mg of PO_4.

Procedure

Take 50 ml of water sample in a flask and add 2-3 drops of phenolphthalein solution. If a pink color appears, add standard sulphuric acid (0.02 N) drop by drop, until the colour disappears. Add 2 ml of ammonium molybdate solution with thorough mixing after each addition. Now, add 0.25 ml (5 drops) stannous chloride after thorough mixing. After 10 minutes, measure absorbance of sample 690nm in the spectrophotometer, using the reagent blank as reference solution. Determine concentration of phosphate in sample from a standard phosphate calibration curve.

Calculation

$$\text{Mg PO}_4 \text{ per liter} = \frac{\text{Mg PO}_4 \text{ from Curve} \times 1000}{\text{ml of Sample Used}}$$

Standard Curve for Phosphate

Take different volumes (1 to 10 ml) of standard phosphate solution in different beakers and dilute (each) to 50 ml with distilled water. Each ml of standard solution contains 0.05 mg of PO_4. So it gives a range of 0.05 to 0.5 mg of PO_4. Add 2 ml ammonium molybdate solution and 0.25 ml stannous chloride solution with thorough mixing, in each sample. Measure the absorbance of each sample at 690 nm in the spectrophotometer, using the reagents blank as the reference solution. Draw curve between measured optical densities and known amounts of PO_4 in mg for them.

3.9 Estimation of Nitrate of Water (Phenoldisulphonic Acid) Method

Nitrate-nitrogen is the most highly oxidized state of the element found in water. In most natural waters, it is also the commonest state. It is brought in aquatic body by the bacterial oxidation of atmospheric nitrogen and by decomposition of organic matter in the watershed.

Apparatus Required

- ☆ Spectrophotometer
- ☆ Glassware

Reagents Required

Silver Sulphate Solution

Dissolve 4.40g Ag SO_4, free from nitrate in distilled water and dilute to one liter.

Phenoldisulphonic Acid Reagent

Dissolve 25 g pure white phenol in 150 ml conc. H_2SO_4. Add 75 ml fuming H_2SO_4 (15 per cent free SO_3), stir well and heat four hours on a hot water bath.

KOH (12N)

Dissolve 673 g KOH in distilled water and dilute to 1 liter.

Ammonium Hydroxide Concentrated

Stock Nitrate Solution

Dissolve 0.722 of anhydrous KNO_3 and dilute to 1.0 liter with distilled water. This solution contains 100mg N per liter.

Standard Nitrate Solution

Evaporate 50 ml (which contains 5 mg of N) to dryness on the hot water bath, dissolve the residue by rubbing with phenoldisulphonic acid reagent and dilute to 500 ml with distilled water. One ml of this standard solution = 0.01 mg N = 0.0443 ml nitrate ion.

Procedure

Take 50 or 100 ml of water sample and determine the chloride contents of water and treat the sample with an equivalent amount of standard silver sulphate solution. Remove the precipitated chloride, either by centrifugation or by filtering coagulating the silver chloride by heat, if necessary. Evaporate the solution to dryness over a hot water bath; rub the residue thoroughly with 2 ml phenoldisulphonic acid reagent to ensure solution of all solids. If need be heat on hot water bath for a short time to dissolve entire residue. Dilute with 20 ml distilled water.

Add with stirring about 6 to 7 ml conc. NH_4OH or about 5 to 6 ml 12N KOH, until maximum color develops. Filter any resulting flocculent hydroxide from the coloured solution. Transfer the filtrate to 50 or 100 ml volumetric flask, dilute to the mark and mix. Measure absorbance of sample at 400-425 nm in the spectrophotometer, using the reagent blank as the reference solution, determine concentration of nitrate in the sample from standard titrate calibration curve.

Calculation

$$\text{mg of N per liter} = \frac{\text{mg N from Standard Curve} \times 1000}{\text{ml of Sample}}$$

Standard Curve for Nitrate

Take different volumes of standard nitrate solutions (these are 1 to 10 ml) in different beakers and dilute to 50 ml (each sample). The standard nitrate solution contains 0.01mg N per ml or 0.0443 mg nitrate ion, so it gives a range of 0.01 to 0.1 mg N. Estimate nitrogen nitrate by above described procedure in each sample. Measure the absorbance of each sample at 400-435 nm in the spectrophotometer, using the reagent blank as the reference solution. Draw curve between measured optical densities and known amount of N or nitrate for them.

3.10 Estimation of Hardness of Water

Water that contains salts of calcium and magnesium and to a lesser extent, other polyvalent metals such as iron, aluminium and manganese, is referred to as hard water. Hardness is now best defined as a characteristic of water representing the total concentration of calcium and magnesium ions expressed as milligrams of $CaCO_3$ per liter. When other ions are present in insignificant amount, the hardness will be equal to or less than the sum of the carbonate and bicarbonate alkalinities and is termed as carbonate hardness. If the hardness exceeds the sum of these alkalinities, the presence of other ions is indicated, and the hardness is assessed as non-carbonate hardness.

Principle

The Criochrome black –I indicator, used in this titration, produces a red colour in the presence of calcium and magnesium. The disodium salt of EDTA forms a stable, colourless complex with these ions, effectively removing them from solution. Thus by titration of the sample, containing the indicator with EDTA, the calcium or any magnesium ions are quantitatively removed. When ions are removed, the indicator changes to a bright blue colour.

Apparatus Required

- ☆ Conical flasks
- ☆ Burette and pipettes.

Reagents Required

- ☆ Ammonia buffer solution
- ☆ Criochrome black indicator
- ☆ Standard EDTA titrant (0.01N)
- ☆ Dissolve 0.3723g Na_2 EDTA, dihydrate in distilled water and dilute to 100 ml. Check by titrating against a standard calcium solution.

  1.00 ml = 1.00 mg $CaCO_3$ = 0.4008 mg Ca

Procedure

Take 50ml water sample and add 1 to 2 ml of ammonia buffer solution to bring pH to 10.0 or 10.1. Add approximately 0.1 g indicator powder. Titrate with standard EDTA solution, continuously, till the colour turns red to violet or bright blue. This indicates the end point. Record the volume of used titrant.

Calculation

$$\text{EDTA hardness as mg } CaCO_3 \text{ per liter} = \frac{\text{Used Vol. of Titrant} \times A \times B}{\text{Vol. of Sample}}$$

where,

A = mg of $CaCO_3$ equivalent to 1.0 ml EDTA Titrant = 1 mg A =1

B = 1000 ml

3.11 Estimation of Calcium of Water

Calcium can be leached from practically all rocks, but is much more prevalent in water from regions with deposits of limestone, dolomite and gypsum. Regions where granite or silicious sand predominate have very low calcium levels in the waters. Waters with a concentration of 10mg of less per liter are usually oligotrophic, while waters with 25mg or more per liter are usually distantly autotrophic.

Principle

In estimation of calcium, the EDTA titrant is used. EDTA solution reacts with both calcium and magnesium. To determine calcium only, the pH is made sufficiently high (by using NaOH buffer) so that the magnesium precipitates as magnesium hydroxide, and an indicator (Murexide), specific for calcium is used.

Apparatus Required

- ☆ Conical flasks
- ☆ Burette and Pipette

Reagents Required

- ☆ Sodium hydroxide buffer (1N)
- ☆ Murexide indicator
- ☆ Standard EDTA titrant

Procedure

Take 50 ml of water sample and add 1-2 ml of sodium hydroxide solution to produce a pH of 13-14. Add about 0.2 g indicator powder. It gives pink colour to the sample. Titrate with standard EDTA with continuous stirring, till the pink colour turns to purple or light blue indicates the end point. Record the vol. of used titrant in ml.

Calculation

$$\text{mg Ca per liter} = \frac{\text{Used Vol. of Titrant} \times A \times B}{\text{Vol. of Sample}}$$

where,

A = 0.4008 (represents that when 1 ml of titrant used, 0.4008 mg of Ca is present)

B = 1000 ml.

3.12 Estimation of Magnesium of Water

Magnesium in natural water comes mainly from leaching carbonate rocks, in areas where maganese concentrations in water often range from 5 to 50 mg per liter. Magnesium is related to water hardness in the same manner as calcium, and it is also an essential nutrient for plant growth and development.

Procedure

There are several analytical methods for magnesium, but when both calcium concentration and hardness are known, magnesium concentration may be calculated from them. Milliequivalents of hardness per liter calculated from milligrams of hardness per liter. The milliequivalents of calcium per liter are subtracted from this and the difference is multiplied by the equivalent weight of magnesium. This gives magnesium in milligrams per liter.

Calculation

meq hardness per liter = mg hardness per liter A

meq Ca^{++} per liter = mg Ca^{++} per liter B

mg Mg^{++} per liter = 12.16 (meq hardness / liter)

where,

A = 0.01998

B = 0.0499

3.13 Estimation of Sodium and Potassium of Water (Flame Emission Spectrophotometer Method)

Principle

Light is emitted by Na^{++} and K^{+} when these elements are excited in a flame, which is measured by the flame photometer (K: 79nm; Na; 589 nm).

Procedure

Follow the instructions provided by the manufacturer. The samples must be free of solids. Prepare a calibration curve. This is often appreciably non-linear. In the case of determination of Na^{+}, dilute the sample until the Na^{+} content is lower than 5mg/1.

Interferences

The nature and extent of interferences depend very much on the characteristics of the flame, and hence on the particular instrument used. Internal standard should therefore be used. No detailed guidance can therefore be given, but it may be noted that in some cases mutual interference of Na^+ and K^+ may be serious. In this case, make a K^+ calibration curve using K^+ solutions, containing the same concentration of Na^+ as is present in the sample and reciprocally for Na^+. Anion interferences (particularly PO_4^{3-} and SO_4^{2}) may also be troublesome.

3.14 Estimation of Sulphate of Water (Turbidometric Method)

In many natural waters, sulphate is the second most common anion being derived from most sedimentary rocks. In lakes, sulphur is cyclic and involves organically reduced form as well as the common free SO_4^- ion.

Principle

Sulfate ion is precipitated with barium ion in acid solution in such a manner as to form barium sulfate crystals of uniform size. Light transmitted by the turbid solution is measured with a photoelectric colorimeter and sulfate ion concentration is read from a standard curve. (Colorimetric equipment –420 nm).

Apparatus Required

- ☆ Spectrophotometer
- ☆ Glassware

Reagents Required

Acid salt solution

Dissolve 240g NaCl in approximately 900 ml distilled water. Add 20 ml conc. HCl and dilute to 1 liter.

Barium Chloride Crystals

20 to 30 mesh size

Standard Sulphuric Acid Solution (0.02N)

1.0 ml diluted to 50 ml will give a solution of 19.2 mg/1.

Procedure

Measure 50 ml sample into a 250 ml flask. If the sample contains more than 60mg/l sulfate, an oblique of portion should be taken and distilled water added to make 50 ml and add exactly 10 ml acid solution and mix it. If the sample is not perfectly clear and colourless at this point, the apparent sulphate content should be measured photometrically and this value subtracted from the determined sulphate concentration. Add one spoonful $BaCl_2.2H_2O$ crystals and start magnetic stirrer, the time of stirring should be constant within 2 sec. Although the total time interval may be any convenient time from 45 sec. to 3 min. One minute is very satisfactory. Allow to stand for an additional 4 min exactly and transfer the resultant suspension to the colorimeter and read immediately. Then read sulfate concentration on a curve prepared from solutions of known sulfate concentrations. Standard 0.02 N sulfuric acid is a convenient solution to use in varying dilutions for preparing the standard curve; 1 ml diluted to 50 ml will give a solution of 19.2 mg/1. The standard curve is not a straight line on semi logarithmic paper, but generally will have a point of infection so that it should be prepared from at least 10 dilutions ranging from 5 to 60 mg/l sulfate.

Calculation

The calibration curve may be plotted in terms of SO_4 concentration, based on a 50 ml sample, versus per cent absorbance. The concentration is then read from the curve. If a different amount of sample was taken, multiply by the appropriate factor.

$$MgSO_4 \text{ per liter} = \frac{MgSO_4 \text{ in Sample} \times 1000}{\text{ml of Sample}}$$

$$= MgSO_4 \text{ in sample} \quad 20$$

3.15 Estimation of pH of Bottom Soils of Fish Ponds

Soil Analysis

Determination of Soil Reaction (pH)

- ☆ Dry 100g of pond's bottom soil in sheds.
- ☆ Take a test tube
- ☆ Add 5 ml of distilled water in the test tube.

☆ Add 2 gm (or ½ teaspoonful) of soil to the test tube.

☆ Add 0.5 to 1 gm of Barium sulphate X-ray quality to the test tube.

☆ Stand this test tube for 20 min with occasional shakings.

☆ Add 5 drops of indicator and shake the contents thoroughly. Allow the soil to settle down completely.

☆ Compare the colour of the upper liquid with the colour chart.

If the result of the above test comes near neutral (7.0) for pH (slightly acidic) then the following procedure should be adopted.

☆ Take a test tube.

☆ Add 5 ml of distilled water.

☆ Add 2 gm (or ½ teaspoonful) of soil to the test tube.

☆ Add 0.5 to 1g of Barium sulphate of X-ray quality to this test tube.

☆ Stand this test tube for 20 min with occasional shakings.

☆ After 20 minutes, add 5 drops of indicator and shake the contents thoroughly and allow the soil to settle down completely. Compare the colour of the upper liquid with the standard chart.

Exercise No. 4

Estimation of Productivities of Water Bodies

4.1 Estimation of Plankton Density of Water

Apparatus

- ☆ Micrometers
- ☆ Microscope
- ☆ Slides
- ☆ Cover glass (22 22 mm or 18 18 mm)

Procedure

4.1a Determination of Area of One Vision Field

First of all, set up the oculometer into eyepiece of the microscope and measure how many divisions of oculometer cover the total area. The value of one division of oculometer is known (1 division of oculometer = 0.015 mm), so total area can be determined by multiplying the total number of divisions of oculometer with 0.015 mm. The area of vision field is not in square, it is circular, so by using the following formula, the total area of one vision field can be determined.

Total area of vision field (in sq. mm) $= \pi r^2$

where,

$\pi = 3.14$; r = radius

$$r = \frac{A}{2}$$

where,

A = Value of divisions of oculometer (in mm) covered by one vision field.

4.1b Determination of Total Number of Vision Fields in One Transect

Take any material on a slide for marking and put a cover glass (22 mm   22 mm or 18 mm   18 mm) on the material. Place the slide under microscope and start counting the number of vision fields from one corner to the other corner of cover glass. By this way, take 3 or 4 measurements and calculate mean value of number of vision fields for one transect.

4.1c Determination of Total Area of One Transect

Total area of one vision field is known (from step a) and the total number of vision field since one transect are also known (from Step b), so total area of one transect can be determined by following calculation.

Total area of one transect (in mm^2) = X   Y

where,

X = Total number of vision fields in one transect

Y = Total area of one vision field

4.2 Estimation of Plankton of Water

Take 40 liter of water sample and filter it through plankton net. Collect the sample from tube of plankton net and make it up to 50 ml. (So this 50 ml of sample represents the 40 liter of water sample). Take 1 ml (about 2 drops) of this sample on slide and cover with cover glass (22 mm   22 mm or 18 mm   18mm). Place the slide under microscope and count the planktons for 5 to 6 different transects and calculate the mean value (No. of plankton) for one transect. Now the total area of one transect (5.1c), total number of planktons in one transect (5.1b) and total area of cover glass (22mm   22 mm) or (18 mm   18 mm) are known, so total number of planktons under cover glass (or per ml, because 1 ml of material was taken on slide) can be determined by the following formula:

$$\text{No. of Planktons (per ml) 'a'} = \frac{\text{Area of Cover Glass}}{\text{Area of One Transect}}$$

where,

- ☆ Area of cover glass = (22mm   22mm or 18mm   18mm)
- ☆ Area of one transect = from 5.1c
- ☆ a = Mean value of No. of plankton for one transect.
- ☆ By above calculation, the number of planktons per ml can be determined and with this value the number of planktons for 50 ml sample (which represents 40 liter of sample) can be determined by following calculation.

  No. of plankton in 40 liter sample or 50 ml = No. of plankton per ml   50
- ☆ By above calculation, the number of planktons in 40 liter can be determined and with this value the number of planktons per liter can be determined.

$$\text{No. of Planktonsper liter} = \frac{\text{No. of Planktons in 40 Liter}}{40}$$

Result = Plankton per liter

4.3 Estimation of Primary Productivity of a Water Body

4.3a Primary Productivity of Water

The primary productivity of an ecosystem is defined as the rate at which radiant energy is stored during the photosynthesis process by a producer organism into the organic substances, which are used as food material.

Primary productivity can be determined by multiplying the rate of photosynthesis by conversion factor for carbon (0.375)

Net primary productivity (NPP) = NP   0.375........mg $C/m^3/3h$

Gross primary productivity (GPP) = GP   0.375- - - -mg $C/m^3/3h$

where,

NP = Net photosynthesis

GP = Gross photosynthesis

4.3b Gross Primary Productivity (GPP) of Water

This is the total rate of photosynthesis including organic material lost by respiration during the period.

GPP = Total new organic matter + Respiratory loss

4.3c Net Primary Productivity (NPP) of Water

This is rate of storage of organic matter in the plant in excess of that lost by respiration during the period.

NPP = New organic matter – Respiratory loss

Principle

Plants produce new organic carbon by the process of photosynthesis. In this process the oxygen is released by plants. So by estimating the oxygen which is released by plants in photosynthesis, the primary productivity can be estimated.

Method

In this method, light and dark bottles are used. For estimation of primary productivity, first the initial dissolved oxygen is determined, then (after a definite time) the dissolved oxygen of light and dark bottles is determined. In light bottle, the level of dissolved oxygen (L) increases then initial level (I), because in this bottle photosynthesis takes place. The increasing value of dissolved oxygen in light bottle can be determined by subtracting initial value of dissolved oxygen. This value is multiplied by conversion factor for carbon (0.375). So this gives net primary productivity (NPP) because the value of respiratory loss (R) can be determined by estimating the dissolved oxygen (OD) in the dark bottle. In dark bottle the value of dissolved oxygen decreases (Because in this bottle, photosynthesis does not take place, so the O_2 is not released but the respiration takes place). The value of dissolved oxygen (D) in dark bottle is subtracted from initial value of dissolved oxygen (I). It gives value of (I) respiration (R). If this value (R) is added to value of net primary productivity, it gives the value of gross primary productivity (GPP).

Apparatus Required

- ☆ Light and dark or standard brown bottle.
- ☆ Rope or other means to suspend pairs of bottles in lake.
- ☆ Other water analysis apparatus as described under dissolved oxygen.

Reagents Required

- ☆ As per dissolved oxygen method.

Procedure

Select station for experiment. Take water sample in light and dark bottles and suspend the bottles in water body by rope or other means. Note the initial time. Take one more water sample at the same time in which estimate initial dissolved oxygen (I) by procedure given in estimation of dissolved oxygen. After three hours, take out the bottles from water body and estimate dissolved oxygen in water sample of both bottles.

Calculations

4.3d Determination of Net Photosynthesis (NP) of Water

Net photosynthesis (NP) = L–1- - - -mg/liter/3hrs
(In form of mg. of O_2/I)

where,

L = Value of dissolved oxygen in light bottle

I = Initial value of dissolved oxygen

4.3e Determination of Respiration (R) Rate from Gross Primary Productivity of Water

Respiration (R) = 1–D - - - - mg/liter / 3 hrs.
(In form of mg of O_2/1)

where,

I = Initial value of dissolved oxygen.

D = Value of dissolved oxygen in dark bottle.

4.3f Determination of Rate of Gross Photosynthesis (GP) of Water

Gross photosynthesis (GP) = NP + R - - - - mg/liter/3hrs.
(In form of mg of O_2/liter)

where,

NP = Value of Net photosynthesis

R = Respiration

4.3g Estimation of Chlorophyll Content of Pond Water

Estimation of Chlorophyll 'a'

The following is a general survey method for estimating the concentration of phytoplankton (chlorophyll 'a') in a volume of water. The value of such a data is uncertain since many variables are present.

Apparatus

- ☆ Spectrophotometer
- ☆ Rubber stoppers or corks
- ☆ Test tubes
- ☆ Membrane filter unit with 0.8 μm, 47mm filters.
- ☆ Centrifuge with stopper glass centrifuge tubes.

Reagents

Ninety per cent alkalized acetone. Using reagent grade acetone and distilled water prepare a 90 per cent solution. Add about one spoonful of powdered magnesium carbonate per liters of acetone solution.

Procedure

Take water sample at several random locations over the lake and at several depths down to limit of photonic zone. Usually one liters samples will be adequate. Filter a portion of well-shaken sample through 0.8 μm, 47 mm membrane filter. Roll up or fold filter with planktons inside. Place filter in bottom of dry test tube. Measure volume of water filtered. Save remainder of unfiltered sample. Add 20 ml of 90 per cent alkalized acetone to sample tube containing filter and planktons. Stopper the tube with an acetone-leached stopper. Shake until filter is dissolved. Allow extraction to proceed in a dark refrigerator. An overnight extraction is preferable. If any turbidity is apparent in extract, centrifuge extract in stopper tubes. Use a blank of 20ml of acetone containing the same number of filters used for sample, and measure absorbance of sample blank at 663 nm. In acetone extracts the absorbance of this turbidity blank may be appreciable.

Calculation

For grass green extracts:

Chlorophyll 'a' concentration = absorbance 7.5*

(mg per liters of extract)

(*7.5 = is used for 1 inch cuvettes; 13.4 is used for 1cm cuvettes).

Chlorophyll 'a' is the chief chlorophyll measured at 663 nm; however, the bank width of the B & L Spectronic–20 probably also measures some chlorophyll 'a' when, it is present.

$$\text{Chlorophyll 'a' concentration of original lake water (mg/m}^2\text{)} = \frac{\text{Chlorophyll in Extract (mg liter)} \times \text{Extract Volume (ml)}}{\text{Filtered Volume (liters)}}$$

Exercise No. 5

Optimal Range of Various Hydro-Biological Parameters

5.1 Water Analysis of a Freshwater Fish Pond

Sl.No.	*Parameters*	*Optimal Range*
1.	Dissolved oxygen (mg/l)	4-10
2.	BOD (ppm)	<30
3.	pH	7.00-8.5
4.	Carbonates (mg/l)	30-50
5.	Total Alkalinity (mg/l)	60-300
6.	Total Hardness (mg/l)	20-200
7.	Conductivity (mohs/cm)	20-1500
8.	Salinity (ppt)	<5
9.	Phosphorus (mg/l)	0.3-0.5
10.	Total nitrogen (mg/l)	0.05-1.5
11.	TDS (mg/l)	<400
12.	Phytoplankton (No./l)	>500
13.	Zooplankton (No./l)	>500

5.2 Bottom Soil Analysis of a Freshwater Fish Pond

Sl.No.	*Parameters*	*Optimal Range*
1.	Texture	
	(*a*) Sand (per cent)	30-50
	(*b*) Silt (per cent)	15-25
	(*c*) Clay (per cent)	10-25
2.	C/N ratio	10:1-12:1
3.	Moisture (per cent)	35-50
4.	Conductivity (mohs/cm)	<16
5.	pH	7.0 -8.5
6.	Total Alkalinity (mg/100g)	60-230
7.	Total Hardness (mg/100g)	50-300
8.	Nitrate (mg)	0.05-7.0
9.	Organic carbon (per cent)	0.5-2.0
10.	Total nitrogen (mg)	>0.50
11.	Salinity (ppt)	<5

Unit–IV

Exercise No. 1

Introduction to Various Fish Diseases

The intensification and diversification of freshwater aquaculture practices in recent years have led to increased incidence of fish diseases in the country. A similar picture is prevalent with regard to brackish water and mariculture system. Fisheries suffer from a wide variety of diseases of infectious and non infectious diseases. It is, therefore, important to know about their etiology in their natural environment. Abundant literature is available with regard to fish mortalities due to infection by viruses, bacteria, parasites, fungi, nutritional deficiencies, metabolic disorders and faulty environmental conditions like oxygen deficiency, high or very low water temperature, chemical pollutants, over crowding, malnutrition, algal blooms, etc. However, fish mortalities are not the only criterion to evaluate the effect of fish diseases. Even the morbidity which leads to weight losses and poor growth in surviving fish contributes for substantial losses to the farmers. The need for maintaining healthy stock as a prerequisite to enhance fish production levels has been fully appreciated by planners and farmers alike in the recent years.

DNA-based methods have been used in diagnosis and for detection of many economically important viral pathogens of cultured finfish and penaeid shrimp. For finfish, tests have been developed for pathogens such as channel catfish virus (CCV), infectious hematopoietic necrosis virus (IHNV), infectious pancreatic necrosis virus (IPNV), spring viremia of carp virus (SVCV), viral hemorrhagic septicemia virus (VHSV), viral nervous necrosis virus (VNNV) and *Renibacterium salmoninarum* (Muroga, 1997; Plumb, 1997). DNA-based methods for detection of penaeid shrimp viruses are now used routinely in a number of laboratories around the world. DNA-based techniques will have an important role to play in efforts to develop sustainable shrimp culture in Asia and elsewhere. Production facilities in Thailand are currently using PCR techniques to screen shrimp post-larvae for white spot syndrome virus (WSSV). Culturing such larvae in closed (biosecure) or semi-closed culture systems can prevent or minimize viral infections, leading to a viable shrimp industry.

1.1 Viral Diseases

Infectious pancreatic necrosis virus (IPNV) is a highly dangerous virus causing contagious viral disease of young fish of the salmonid species. It is widely reported in Europe, Asia, America and Canada. Significant mortalities can occur from IPN outbreaks in fry. Disease signs include: Pale liver and spleen, stomach and intestine devoid of food, darker skin and trailing faecal casts. IPNV can cause an asymptomatic carrier state in salmonid fish. These carriers can shed the virus and transmit the virus to their progeny and other susceptible species. It is therefore important to detect asymptomatic carriers in order to limit the spread of infection.

Two new methods *i.e.* destructive (Gahlawat *et al.*, 2004a) and non-destructive (Munro *et al.*, 2004; Gahlawat *et al.*, 2004b) were developed for the detection of IPNV carriers in Atlantic salmon (*Salmon salar* L.) and halibut (*Hippoglossus hippoglossus* L.). These methods are easy, inexpensive and very sensitive as compared to OIE (Office International des Epizooties 2000) standard tissue culture and RT-PCR. No information was available on the long-term persistence of IPNV in halibut and the fish were detected carriers by these methods. The destructive methods of viral isolation involve the need to kill the fish to obtain kidney and liver tissue. As halibut are repeat spawners, for their broodstock testing there was a great need to develop a sensitive non-destructive method for isolation of IPNV. Macrophages are phagocytic cells and capable of killing wide range of pathogens. These play a vital role in immune response of fish and hence utilised in developing these methods.

Spring viremia of carp (SVC) is a viral disease of fish, primarily common carp *(Cyprinus carpio)* caused by ssRNA Vesiculovirus (Rhabdoviridae), known as Spring viraemia of carp Virus (SVCV) or *Rhabdovirus carpio* (RVC) (Fijan,1999). It was first described in 1971

(Fijan, 1972). Before the disease was recognised, it was variously called infectious dropsy, infectious ascites, hemorrhagic septicemia, or rubella. In Europe, the disease has had substantial impact on the production of carp, with estimated losses of 10–15 percent of 1–year–old carp or about 4,000 tons annually. In some cases, mortality rates of young carp can reach 70 per cent. This impact of the disease has led to its listing by the Office International des Epizooties (OIE) as notifiable. In Europe, it was quite some time before SVCV was adopted as a viral disease; many people thought it was caused by bacteria especially *A. hydrophilla*.

The disease has been reported to occur in most European countries, the Middle East, Australia and Asia. Recently the disease has been reported from Brazil and the United States of America (Dixon, Peter EU workshop, CEFAS). The virus has also been isolated in United Kingdom from fish exported from China, although the disease has not been reported from China (Dixon, Peter EU workshop, CEFAS) and India (Asia-Pacific Quarterly Aquatic, Animal Disease Reporting System 1999–2000)

First signs of the disease may be a change in behavioral patterns of the fish. Fish may congregate in slow–flowing water, near pond banks, or lie on the bottom. Over time, the rate of respiration will decrease, as will reaction to stimulation and swimming speed. As the disease progresses, the fish becomes sluggish and may swim and lie on their sides.

Externally, the fish can exhibit a number of non–specific physical signs including darkening of the skin, swollen abdomen, exophthalmia (pop–eye), haemorrhages in the skin, gills and anterior eye chamber, anaemia and pale gills, and a protruding vent. Internally, the signs are dominated by building up of fluid (oedema) in all organs and in the body cavity, haemorrhages in the swim bladder, and inflammation of the intestines (Fijan, 1999).

Koi Herpes Virus

In 1988 a disease emerged that caused mass mortalities in populations of cultured common carp (*Cyprinus carpio carpio*) and koi carp (*Cyprinus carpiokoi)*. Devastating losses occurred on farms culturing carp in Israel and Germany and outbreaks of disease with similar etiology occurred in koi carp in USA, England, Germany and Netherlands. A herpesvirus was isolated from carp during many of these outbreaks that was later shown, in transmission studies, to be the cause of the disease. Outbreaks of the disease tend to be seasonal and occur at water temperature between 18 and 22°C.

The disease is highly infectious and the most prominent clinical sign is discoloration of the gills. Other signs include lethargy combined with short bouts of hyperactivity, endothalmia (sunken eyes), pale patches on the skin and an increase in mucus production on the skin and gills. Microscopic examination reveals severe hyperplasia and necrosis of the gill epithelium, and nuclear hypertrophy and margination of chromatin can be observed in some epithelial cells.

1.2 Bacterial Disease

Innumerable disease are caused in fishes due to bacterial pathogens and several of them are reported in Indian literature. They occur in nursery, rearing and grow out ponds causing serious concern to fish farms. Some of the important bacterial pathogens are *Aeromonas hydrophila, A. salmonicida, Pseudomonas fluorescens, Pseudomonas* sp., *Flexibacter columnaris, Edwarsiella tarda, Vibrio alginolyticus, V. anguillarum* and *V. parahaemolyticus* which have been identified as the most commonly occurring bacterial agents of fish disease. Symptomatology, clinical pathology and anatomic pathology have been carefully correlated to arrive at the diagnosis of the diseases caused by these organisms.

Fin and tall rot in young fish are reported to be due to a mixed infection of *A. hydrophila* and *Pseudomonas fluorescens*. Bath treatment in 1:2000 copper sulphate for 1-2 minutes or painting of concentrated copper sulphate solution in the affected fish act as a good control measure. However, now a days, iodine solutions and antibiotics are used to contain fin and tail rot.

Dropsy is another important fish disease in India where rohu, catla and mrigla are affected mostly in the late winter and have been reported in composite fish culture ponds. The epidermis and body cavities get filled with fluid and scales protrude out from their pockets leading to severe anaemic condition. Terramycin in the feed for about 2 weeks is reported to have good effect. A mixed infection of *A. hydrophila* and *Myxosporidian* parasite is often reported in cases of *Catla catla*. Some workers in the country have attributed cases of dropsy to *A. hydrophila* in conjunction with malnutrition.

Figure 15: Fin Rot in *Catla catla*

Another bacterial disease in the epidemic form has been the pop-eye-disease the causative organism of which is identified as *A. liquifaciens* and *E. tarda* the cornea of the eyes get vascularized leading to opacity and complete necrosis resulting into death of the fish Corneal opacity in silver carp (*Hypophthaimichrhys molitrix* Val) and *Channa narulius* due to grain-positive bacterium, *Staphylococus aureus* has also been reported. Chloromycetin bath @5-10 mg/litre of water has been found effective in controlling the disease at an early stage by some workers. Disinfecting the environment with $KMnO_4$ at a dose of 0.1 ppm followed by liming @300 ppm check the disease considerably.

Myxobacterial organisms have been attributed to cause a condition known as gill hyperplasia syndrome in common carp (*Cyprinus carpio*). The infectious nature of the disease has been proved by transmitting the disease in Asiatic major carp species through contact. Ulcerative disease in catla has been reported by many workers showing bilateral ulcerations of the opercular bones and cranium and *A. hydrophila* was isolated from the lesions. Another disease of similar nature causing external lesions over the surface of the body has been described by several workers. The causative organism has been identified as *Flexibacter columnaris*. Topical application of potassium permanganate or short bath in 500 ppm of $KMnO_4$ has been found to be very effective in completely curing the disease.

A septicaemic disease called Edwardsiellosis has been reported affecting brood fish population. *Edwardsiella tarda* has been isolated from the diseases fish showing anaemia, cutaneous lesions and gas filled abscesses in the muscles. With considerable morbidity, even large scale mortality of spawn of Indian major carps has been reported in organized hatcheries. The young ones showed anatomical deformities and greyish with discoloration of the body within 48 hours post hatching, and died within 72 hours. Although treatment with lodophor has been found to be effective, water quality improvement in the hatchery is the most essential component for keeping this disease away.

Epizootic Ulcerative Syndrome (EUS) has become a matter of tremendous concern not only among the fishermen and fish farmers, but also among general public, enterpreneurs, administrators and planners. Initially, the epizootics were mainly reported from the North-Eastern states of the country in the year 1988. Gradually, it spread throughout the country and has engulfed the entire length and breadth of the country barring only few states like the Punjab, Haryana, Jammu and Kasmir, Himachal Pradesh and Gujurat. The

Figure 16: Epizootic Ulcerative Syndrome in *Cirrhinus mrigala*

disease has been reported not only in the freshwater fishes, but also brackish water fishes in Orissa, Kerala West Bengal, Tamil Nadu, Karnataka and Goa. One common feature of the disease is that it initially affects the bottom dwelling species like murrels followed by catfishes and weed fishes. Subsequently other fishes including Indian major carps are affected. Although the extent of economic loss has not been properly quantified, loss incurred by the fisheries sector due to this particular disease runs to several crores of rupees.

1.3 Protozoan Disease

Myxosporidia are one of the most important group of pathogens capable of producing disease in fish causing heavy loss in the juveniles. Several species of this pathogen have been reported from various parts of the country causing heavy losses in intensive fish culture and fish seed rearing due to emaciation and asphyxiation. Myxobolus cysts of varying sizes have been reported on the gills and kidney of *Catla catla*. Diagnosis of the disease can be made on the basis of gross appearance of the pin head sized greyish cysts and large number of myxosporidian spores under the microscope.

The other group of protozoa infecting fish belong to cillophora. The carp fry and fingerlings sometimes carry heavy infection of the gills and skin with *Tricodina indica* when their is excessive mucus secretion on the gills which hampers respiration and the fish start surfacing. Some species of *Tricodina* occur inside the urinary bladder. Fish affected with tricodiniasis eventually turn sluggish. A short bath in 2-3 per cent salt solution for 5 minutes can control the infection effectively

Whirling disease in *Cirrhirus mrigala* (Ham) has been attributed to a protozoan parasite, *Myxosoma cerebra*. The gross changes are characterized by deformed spinal column as reported in European and American salmonids suffering from whining disease. Ichthyophthiriasis or "white spot" disease has been reported causing large scale mortality and rearing ponds in different pans of the country. Presence of numerous pin head shaped white spots on the skin, fins and gills are common diagnostic features.

1.4 Helminthic Parasites

Fishes are infected on their gills and skin by many species of monogenetic and digenetic termatodes and cestodes. The important monogenetic species are *Gyrodactylus* and *Daclylogyrus*. They remain attached to the skin or gills by the help of anchors which cause wounds. Their presence cause hypersecretion of mucous and hemorrhage from blood capillaries. These wounds later get infected with pathogenic bacteria and fungi. Alternate bath in 1:2000 acetic acid solution followed by bath in 2 per cent sodium chloride solution has been found to be effective. Some workers have advocated dip treatment in 5 per cent salt solution followed by bath in 1:5000 formalin.

The digenetic termatodes generally infect the alimentary canal but are rarely harmful. However, at the metacercarial stage, they are harmful as they get encysted in any part of the body, *i.e.* skin, muscle, abdominal cavity, liver, eyes and brain. In these cases, fish serve as intermediate hosts while the final host is generally a fish-eating bird. In many carp fingerlings; the black spot disease is due to metacercaria of diplostornatidae. Catla and silver carp are the most susceptible species

Cestodes are also internal parasites in alimentary canal, with higher infestations observed in catfishes than in carp. The infected fish show softening of the muscle tissue probably due to accumulation of water. In cases of heavy infestations, the abdomen bursts resulting in death of the fish. Acanthocephala have been recorded from major carps and catfishes in India, but the life history or biology of the parasites has not been studied in detail. The parasite causes lesions in the intestinal wall by its proboscis.

Parasitic copepoda are external parasites infecting skin or gills of bucco-pharyngeal cavity. Among them, species of two genera *Lernaea* and *Ergasilus* are reported from Indian carps *Lernae* is reported to infect fingerlings and adults of major carps, especially catla. The parasite penetrates the muscles and reaches the blood vessels as it feeds on the host blood, rendering the fish anaemic and weak. *Ergasilus* sp. Infect the gill lamellae. They attach themselves to the gills by the help of their book like second antenna and cause hemorrhage in the gill tissue.

1.5 Crustacean Parasites

Heavy mortality due to Argulosis or fish lice infestation have been reported in Indian major carps. Several species of *Argulus* have been reported to infect cultured fishes. These parasites inhabit the skin, fins and gills of fish host and cause extensive pathological lesions in the skin showing circular haemorrhagic patches which become ulcerated. Mucous cells proliferate and copious mucous is produced. These lesions promote secondary infections like fungi and bacteria. Rohu has been observed to be the most susceptible species among the carps.

Bath in potassium permanganate @ 500 ppm for one minute once a week in combination of malathion treatment is helpful in the eradication of the parasite.

1.6 Fungal Disease

Saprolegnia parasitica in the fry and fingerlings of major carps is characterized by a white to brown cotton like growth consisting of colonies of mycellium and filaments which appear as small to large patches on various parts of the body like fins, gills, mouth, eyes or muscle. The infection starts due to netting injury and over-crowding or lesions caused by other diseases. *Saprolegnia* often infects the fertilized eggs. They attack the dead eggs first and then spread to the viable living eggs resulting in their spoilage. Infection due to *Aphanomyces pisci* is another fungal disease where the scales become black and later fall off. The hyphae of this species are not visible like those of *Saprolegnia*.

Branchiomyces infection causes "gill rot" in fish. These pathogens invade the blood vessels of gills causing necrosis of the surrounding tissues also. Apart from these, *Aspergilus flavas, A. ochraceous* and Fusiformis momiliforme have been reported from some of the virus affected specimens.

Several different treatment measures for fungal infection in fish have been formulated, which include dip treatment in 3 per cent common salt solution, 1:2000 copper sulphate solution or 1:1000 potassium permanganate solution for 5-10 minutes. A flush treatment for 1 hour in malachite green (0.1 ppm) is found to be quit effective in curbing fungal infection.

1.7 Centres in the Country Involved in R&D Program on Aquatic Health Management

The institute under Indian Council of Agriculture Research (ICAR) such as the Central Institute of Freshwater Aquaculture (CIFA), Bhubaneswar, the Central Institute of Brackish water

Aquaculture (CIBA), Madras, the Central Institute of Capture Fisheries Research Institute (CICERI), Barrackpore, and Central Marine Fisheries Research Institute (CMFRI), Cochin are the research organizations on which lies the responsibility of research and training of aquatic health management with particular reference to freshwater, brackish water and marine finfish and shellfish. These institute have developed capabilities of their own on disease diagnosis in fish and prawns and organizing training for fisheries officials, teacher and students through effective human resource development and establishment of laboratory facilities. These are the nodal organizations on which the state and central government depends for research & technical information related to fish health and disease control.

Besides these institutions, there are other university college *e.g.* Department of Zoology & Aquaculture, CCS Haryana Agricultural University, Hisar; Fisheries College, Mangalore under Karnataka Agricultural University, Veterinary College, Kerala Agricultural University, Trichur; The School of Life Sciences, Bharatidasan University, Tiruchirapalli; The P.G. Dept. of Zoology, American College, Madurai; The P.G. Dept. of Zoology, Andhra University, Visakhapatanam; Dept. of Fisheries Science, Barkatullah University, Bhopal etc., are some of the organizations which pursue R&D programs on various aspects of aquatic monitoring and management.

Mostly the R & D programs of the country are founded by organizations like the JCAR, Department of Biotechnology, D.S.T.., NAM', NABARD, CSIR etc., besides foreign agencies *e.g.* FAO/UNDP, World Bank, USAID, ODA etc. After the outbreak of two serious diseases in finfish and shrimps in the country, the World Bank has recently come forward with the intention to strengthen assistance. This will help modernization of the laboratories and training of personnel.

1.8 Aquatic Environment Management

Aquatic environment research that, till a few years back was confined to analysis of physico-chemical practices, has assumed great importance and wide implications in the work of intensification and diversification of aquaculture practices. The researchers have make efforts in widening the spectrum of studies on aquatic environment, and both aspects of productivity and pollution are being addressed to with due emphasis on recycling of organic matter into aquaculture. The studies have been made on nutrient regime and productivity levels in fish ponds under different management. Some of the recent contributions in the field of aquatic environment areas are as follows:

1. Quantification of primary production, community respiration, sediments oxygen consumption with reference to bacterial and macro-invertebrate respiration in rural and farm ponds.
2. Emumeration of heterotrophic and other bacterial groups involved in nutrient cycles in fish ponds under different management practices and estimation of decomposition potentials of fish ponds.
3. Quantification of decomposition rates and nutrient liberation rates of organic manures and selected organic substrates along with the associated bacterial populations and oxygen consumption rates
4. Quantification of detritus and plankton levels in nursery, rearing and stocking ponds and demonstration of significance of heterotrophic food chain.
5. Utilization of processed water hyacinth and other lignocellulosic wastes in carp culture as fees and manure with studies on associated bacterial populations.
6. Studies on bacterial populations and nutrient regime of biogas slurry and its utilization in different aquaculture practices like carp nursery, rearing and stocking ponds and duckweed culture as a nutrient input. Provide a total package of practices for use of biogas slurry, to produce 3-4 tones fish/ha/yr. with biogas slurry applied @ 30-45 tones/ha/yr. Slurry also found useful as a carp feed ingredient and in preventing of pond seepage.
7. Demonstration growth-promoting effects of alpha-tocopherol acetate in algal cultures and pond primary production rates, and trophic significance of aquatic microbial communities in carp culture. Evaluated the use of spirulina also as an ancillary practice in fish culture.

8. Quantification of biological nitrogen fixation levels in water and sediment media of fish ponds under different management practices (3-5 kg. N/ha-m/yr)? being the first estimates in the country. Further, elaborated the bacterial communities, seasonal and spatial variations in nitrogen fixation rates, factors influencing the process, etc.
9. Evaluated biofertilization measures with free-living bacteria (*Azotobacter* sp.), cyanobacteria and *Azolla* application. Standardize the application of *Azolla* in carp culture ponds as a biofertilizer @ 40 tones/ha/yr, to provide the full complement of N-P-K and organic matter in carp culture, along with an *Azolla* culture system suitable for aquaculture practices.
10. Quantification of nitrogen and phosphorus budgets and sediment-water nutrient interactions using radioisotopes (C^{14}, P^{32}, N^{15}) and stable isotopes (15N) with nutrient rates by trophic components of fish pond systems.
11. Evaluation of biofilter substrates and nitrification rates for incorporation in hatchery and culture system.
12. Studies on bacterial complexes of carp, catfish and freshwater prawn hatchery system along with amelioration measures.
13. Quantification of energy transfer processes in fish pond ecosystems through ecological efficiencies in different culture systems along with studies on bioenergetic pathways.
14. Quantification of methane emission rates from fish pond ecosystem.

A new dimension of environmental research in wastewater treatment through aquaculture, both long-term researchers on sewage-fed fish culture system at bheries of Calcutta, new line of work pertaining to treatment of domestic waste through duckweed-fish culture is being pursued under the project funded by the Ministry of Environment and Forests, Govt., of India. An aquaculture complex for treating sewage as been established on the river Kathojadi at Matagajpur at Cuttack for treating 1 mld of sewage culture with different duckweeds like *Lemna, Spirodela* and *Azolla* along with carp culture ponds, being the first of its kind in the country.

Realizing the Importance of organic farming in agriculture practices, an Operational Research Project on Processing and utilization of organic waste in aquaculture has been launched in ten States of the country, viz Orissa, West Bengal, Bihar, Madhya Pradesh, Maharashtra, Andhra Prasesh, Uttar Pradesh, Karnataka, Tamil Nadu and Kerala. The project provides for assessment of manurial and feed resources in different agro-climatic zones of the country, formulation of sustainable processing methods and use of biofertilization material in aquaculture.

With the potentials of utilization of a host of organic wastes in aquaculture on one hand and possible environmental effects due to aquaculture operations on other, efforts are being made to formulate eco-friendly aquaculture practices to optimize the production level. The Institutes like the Central Institute of Freshwater Aquaculture, Central Inland Capture Fisheries Research Institute, Central Institute of Brackish Water Aquaculture, National Environmental Engineering Research Institute and a host of Universities and Fisheries colleges are involved in this endeavour in the country.

Exercise No. 2

Immunological Techniques

Resistance of animals to infection by bacteria, viruses, fungi and parasites is called 'immunity'. In response to natural infection by pathogens or on deliberate inoculation of antigens, the host body immune system gets activated leading to immune response. Science of immunology originated as a branch of medicine dealing with novel approaches to control infectious diseases of man and animals more than two centuries ago. The immune response against invading pathogens/antigens is of two types: Humoral immune response, characterised by production of immunoglobulins which are called antibodies and Cellular immune response, which is characterised by activation of mainly T lymphocytes. Both types of responses help the individual to fight the pathogenic agents. The activation of immune system does not only cause removal of the antigen but also generates memory. This memory leads to a faster and stronger immune response in the future exposures with the same antigen. Mainly three types of cells (B lymphocytes, T lymphocytes and monocytes) participate in immune response against an antigen.

An antigen (Ag) is defined as any foreign susbtance which is capable of eliciting an immune response in the host. Lately the term 'immunogen' is used more frequently instead of antigen. Macromolecules such as proteins, nucleic acids, carbohydrates and lipids can all act as antigens. Most of the protein antigens contain several antigenic determinants. When such antigens are introduced into animals and human beings, they produce antibodies against each of the antigenic determinants. Serum obtained from such animals will contain a mixture of antibodies with specificity to individual immuno-dominant domains. Such sera are called 'polyclonal sera'.

2.1 Immunological Methods Based on Antigen-Antibody

Precipitation Reaction Based Assays

Combination of molecules of soluble antigen with specific antibodies results into lattice formation and when this complex attains a certain size, it becomes insoluble and precipitates out in the form of a visible band at the site of reaction. The phenomenon of precipitation is called precipitin reaction. When increasing concentration of the antigen are added to a series of tubes that contain a constant amount of the antibodies, varying quantity of precipitae is formed.

Agglutination Based Tests

When a particulate antigen combines with a specific antibody, it results in agglutination for formation of clumps of the reactants. These tests can be performed either on a slide or in a test tube.

Radioimmunoassay (RIA)

This technique enables measurement of picogram (10^{-12} g) of an antigen with relative ease, precision and rapidity by using radioactively labelled antigen. This method is highly sensitive and can thus be employed for accurate micro determination of specific molecules in unfractionated mixtures. The RIA suffers from certain inherent drawbacks such as short half life of the radiolabel, requirement of highly skilled technicians, expensive infrastructure, potential health hazards associated with routine use of radioactive material and the problem of disposal of radioactive waste. The term 'enzyme linked immunosorbent assay' (ELISA) was coined in 1971 by Engval and Perlaman for this enzyme based technique. There are a variety of modifications of this test and these are collectively known as enzyme immunosassays (EIA). The basic feature of this group of assays is that an appropriate enzyme is conjugated with the antibody. It is then incubated with the test sample or the test sample which has been previously treated with primary antibody against the antigen. The labelled antibody specificially combines with the antigen or the antigen-primary antibody complex. The unbound enzyme labelled antibody is removed by washing. The enzyme labelled immunoreactant is then conveniently detected and quantitatively estimated by assaying the enzyme associated with antigen-antibody complex.

Exp. 1: To Produce Polyclonal Antiserum against Bovine Serum Albumin (BSA)

When a 'foreign' substance such as a protein or other macromolecular substance is injected into an animal its immune system gets activated leading to synthesis of antibodies specific to the antigenic determinants (epitopes) present on the antigen and are released in the serum. Since different B lymphocytes recognise individual epitopes, the antibodies thus produced is a mixture of the ones having different specificities as these are produced by different clones of activated B cells. Hence the antiserum thus produced is referred to as polyclonal antiserum. To produce polyclonal antiserum against a particular antigen, a suitable animal is immunised with pure homologous sample of the susbtance. For example, antiserum against BSA can be raised by injecting it into rabbits. Normally antigen is mixed with appropriate adjuvant to elicit a good immune response. A few days (usually 7-21 days) later a 'booster' injection is given to the animal which considerably enhances antibody response. After few days, the blood is collected and allowed to clot. The clot is broken and the cells are removed by centrifugation and the supernatant which represents antibody containing blood serum is aseptically collected and kept at –20 to –70°C for long time storage, at 4°C if it has to be used within a short period.

Reagents

1. Chloroform or anaesthetic ether
2. Fruend's complete and incomplete adjuvants. Available commercially.
3. Bovine serum albumin (BSA): A highly purified preparation of BSA should be free from any contaminating proteins.
4. Normal saline solution: Prepared by dissolving 8.5 g NaCl in water and making its volume to 1 *l* with distilled water.

Methodology

1. Suspend 10 mg/ml BSA antigen in normal saline solution in a clean and sterile beaker. To 6 ml of BSA solution add an equal volume of Freund's complete adjuvant. Make emulsion with the help of syringe by repeatedly drawing and ejecting the mixture through 18 G needle.
2. When a complete emulsion has been formed, inject 1 ml of emulsified BSA intramuscularly (I/M) into each rabbit.
3. Give booster injection (I/M) of the same dose of antigen but now in Freund's incomplete adjuvant after 4-6 weeks of the first injection.
4. Test bleed the animals from ear vein 5-7 days after the booster inoculation and collect the blood sample.
5. Separate the serum by allowing the blood to clot at room temperature. Remove any remaining cellular material from the serum by centrifuging it at 2000 rpm for 20 min at 4°C in refrigerated centrifuge.
6. If the serum contains desired level of antibody titre, kill the animals by cardiac puncture. The bleeding of animals by this route allows collection of large volume of blood and thus higher quantities of polyclonal antiserum.
7. Store the antiserum at 4°C for short term use but at –20 to –70°C if it has to be preserved for long time. The antiserum can also be lyophilised and stored at 4°C for long time use.
8. To avoid contamination, either filter the antiserum through 0.45 um membrane filter or add antimicrobial agents like merthiolate (0.01 per cent) or sodium azide (0.1 per cent).

Exp. 2: Titration of Polyclonal Antiserum by Radial Immunodiffusion Assay

Titre value of the test sample can be determined by immunodiffusion assay. Homologous antigen is incorporated into agarose solution which is then allowed to form a gel. Circular wells of uniform size are cut into the gel to which test serum sample containing the antibody is added. On diffusion of the antibody, a ciruclar precipitation zone is formed around the well whose area is proportional to the concentration of antibody. In other wells, standard solutions containing known amount of the antibody are added. A plot of antibody concentration versus square of diameter of the precipitate ring gives straight line thus enabling preparation of standard curve. From the diameter of the ring formed around the well containaing the test serum sample, concentration of antibody in it can be determined from the standard curve. If antiserum with known concentration of the antibody is not available, titre value of the test antiserum sample can be determined by serial dilution method as described in this experiment.

Reagents

1. Agarose
2. Bovine serum albumin (BSA)
3. Antiserum against BSA
4. Barbitone buffer (pH 7.8): Dissolve 1.5 g NaCl, 1.5 g KCl, 1.25 g sodium barbital, 1.0 g $MgSO_4$ and 100 mg NaN_3 in distilled water. Make the final volume to 500 ml with distilled water.

Methodology

1. Take 2 per cent (w/v) agarose in barbitone buffer and melt it by boiling.
2. Keep the melted agarose at 56°C in water bath.
3. Prepare several dilutions of BSA in barbitone buffer and keep it at 56°C in a water bath.
4. Mix equal volumes of melted agarose with diluted antigen solution and pour the mixture onto a glass slide or a petri plate so that approximately 1 mm thick layer of agarose is formed. Prepare separate plate for each antigen dilution.
5. After solidification of agarose, punch 6 holes into agarose and remove the agarose from the punched holes by suction.
6. Seal bottom of the wells by adding melted agarose.
7. Fill each well by serially diluted anti-BSA antiserum.
8. Place the plates in humidified chambers and incubate at 25°C for 24 h in BOD incubator.
9. Observe the plates and record diameter of the precipitate around the wells.

10. The dilution of antiserum beyond which there is no visible ring around the well is the titre of the antiserum. However, the other way to find out the anti-BSA antiserum titre is by using standard curve obtained by using preparation containing known concentration of the antibody.

Exp. 3: To Detect Antibodies in Sera by Competitive ELISA (cELISA)

In this test, serial dilutions of the test sera are made and allowed to interact with the antigen. Then high titred monoclonal antibody prepared against the purified antigen is added. In the test serum of lower dilutions, the number of antibodies binding to antigen will be much larger and thus monolclonal antibodies would find fewer sites on the antigen to bind. Subsequently, upon addition of enzyme conjugated with anti/mouse IgF, a fewer molecules of this conjugated will bind and thus little colour would only develop. At higher dilution of the antibodies in the test sera get diluted and more sites on the antigen would become available for the monoclonal antibodies to bind and thus the colour developed would be more. Evidently the colour intensity obtained would be inversely proportional to the amount of antibodies in the test serum.

Materials

1. Antigen
2. High titre known serum specific to the antigen produced in any species other than mouse.
3. Low litre known serum specific to the antigen produced in any species other than mouse.
4. Antigen specific monoclonal antibody or standard polyclonal antiserum.
5. Anti-species immunoglobulin conjugate (anti-mouse)-commercially available.
6. Substrate solution: Substrate solution will be prepared as standard norms.
7. Carbonate bicarbonate buffer (pH 9.6 coating buffer): Dissolve 0.15 g sodium chloride, 0.29 g sodium bicarbonate and 0.02 g sodium azide in water and make it to 1 *c* with distilled water.

8. Washing buffer, pH 7.4 (PBST): Prepare a solution comprising 8.0 g sodium chloride, 0.20 g potassium chloride, 1.45 g disodium hydrogen phosphate and 0.2 g potassium dihydrogen phosphate in 1 *l* distilled water. Add Tween-20 to this solution so that its final concentration is 0.05 per cent (v/v).
9. Skimmed milk powder: Prepare 5 per cent (w/v) solution of skimmed milk powder in distilled water.
10. 1 M H_2SO_4.

Methodology

1. Coat each well of ELISA plates with 50 µl of diluted antigen. Dilute the antigen appropriately in coating buffer. Incubate on plate shaker for 1 h at 37°C or over night at 4°C.
2. Wash the plates thrice with washing buffer and tap the plates to remove the residual solution from the wells.
3. Add 50 µl diluted sera in triplicate to the antigen coated wells. Incubate on plate shaker for 1 h at 37°C.
4. Add 50 µl of diluted antigen specific monoclonal antibody (Mab) to each well. The Mab is diluted in PBST + 1 per cent skimmed milk powder. Incubate on the plate shaker for 30 min at 37°C. Wash the plates three times and remove any remaining solution by tapping.
5. Add 50 µl of TMB substrate (as per the instructions in the kit) to each well. Incubate at 25°C for 10 min.
6. Stop the reaction with 50 µl of 1 M H_2SO_4.
7. Record absorbance of the formed colour in each well of plates at 450 nm in ELISA reader.
8. Covert the recorded experimental values into per cent inhibition using the following formula:

% inhibition =

$$\frac{A_{450}\text{ Non - reactor (Negative Serum} - A_{450}\text{ Test Serum)}}{A_{250}\text{ Non - reactor (Negative Serum} - A_{450}\text{ High Titre Serum)}} \times 100$$

The per cent inhibition values greater than 50 are considered as positive while less than 50 as negative for the antibodies in the test sera.

Exp. 4: Qualitative Determination of Antigens by Immunoblotting Technique

Immunoblotting involves transfer of electrophoretically separated proteins of the test sample from the SDS-PAGE gel to a solid support such as nitrocellulose membrane and subsequently probing western blots with specific antibodies. These antibodies would react specifically with antigenic epitopes on the target proteins and thus facilitate in establishing the presence or absence of the antigen under investigation. Thus, immunoblotting combines resolution of SDS-polyacrylamide gel electrophoresis with specificity of immunochemical detection. It is, therefore, very useful in determining the presence of antigen as well as relative molecular weight of polypeptide chain. The proteins identified by immunoblotting are highly denatured. Hence conformational epitopes are lost leading to failure of binding of antibodies specific to conformational epitopes. Therefore, it is advisable to use polyclonal antiserum which contains antibodies to a variety of epitopes. This is the reason why many monoclonal antibodies fail to react in immunoblotting.

Reagents

1. PBS (phosphate buffer saline, pH 7.4).
2. Blocking solution (5 per cent skimmed dry milk powder in PBS, pH 7.4).
3. Hyper immune serum against the antigen.
4. Anti-species horse radish peroxidase (HRPO) conjugate (Sigma Co., USA)
5. Hydrogen peroxide, 30 per cent
6. DAB (3,3′-diaminobenzidine tetrahydrochloride)
7. Substrate solution: Dissolve 6 mg of DAB in 10 ml mM Tris (pH 7.6). Filter through Whatman No. 1 filter paper to remove precipitate. In the filtered substrate solution add 10 ml of 30 per cent H_2O_2 just before use.
8. Nitrocellulose membrane
9. Washing buffer (PBS, pH 7.4)
10. Bovine serum albumin (BSA)

Methodology

1. Blot the polypeptide pattern onto nitrocellulose membrane from the SDS-PAGE gel.
2. Block non-specific binding sites on the nitrocellulose membrane by immersing overnight in blocking solution.
3. Remove the membrane from the blocking solution and wash twice for 5 min in PBS. Add appropriate dilution of hyper immune antiserum raised in guinea pig and diluted in PBS containing 3 per cent (w/v) BSA. Incubate for at least 1 h at room temperature with agitation. Wash the membrane with 4 changes of PBS for 5 min each.
4. Add anti-guinea pig HRPO conjugate diluted in PBS containing 3 per cent (w/v) BSA and incubate for 1 h at room temperature with agitation. Wash the membrane with 4 changes of PBS for 5 min each.
5. For development of immunoblot, incubate the membrane in appropriate amount of substrate solution to which 10 ml of 30 per cent H_2O_2 is added and develop at room temperature with agitation until the sufficiently dark visible bands (brown colour) appear. Stop the reaction by rinsing with PBS.

Precautions

1. Use of appropriate amount of antibody and conjugate solutions and adequate shaking should be done to ensure proper antigen-antibody reaction.
2. The problems of high background and other artifactual stains on the immunoblot can be minimised by effectively blocking the binding sites on the membrane and adequate washing.
3. If background problem still persists, reduce the concentration of labelled antibody. For this, the concentration of labelled antibody may have to be optimized for best resolution without much background staining.
4. Sometimes uneven background may be seen in the immunoblots which may occur due to improper handling

of membrane. Thus it is advisable to use gloves when manipulating the membrane with hands.

2.2 Radioimmunoassay of T_3 and T_4

The technique of radioimmunoassay (RIA) permits the precise measurement of antigens upto pico-gram (10^{-12} g) quantities. Radioimmunoassays have proved to be very useful for the quantification of hormomes in very small volumes of biological samples and eliminate often a need for purification or extraction of the principle to be analysed from samples. It is basically a competitive binding assay in which fixed amounts of antibody and labelled antigen react in the presence of known (standards) or unknown (samples) amounts of the antigen. The amounts by which the binding of labelled antigen to its antibody are competitively inhibited by an increasing amounts of standard antigen preparation are recorded (dose response curve or standard curve). The amount of antigen present in an unknown sample can then be calculated from the standard curve by comparing the binding of labelled antigen. The assay in which both labelled and unlabelled antigens are allowed to compete simultaneously for the limited number of binding sites (limited amount of antibody) is called equilibrium type of RIA. The one in which the unlabelled antigen (standard/sample) is first allowed to bind the antibody (first incubation) and then further incubated with labelled antigen is known as non-equilibrium type of RIA.

In the simplified way, the reactions involved in RIA can be summarised as follows:

		Bound		Free
(*i*) Without competition				
12Ag* + 6 Ab	=	6 Ag* Ab	+	6 Ag*
(12000 cpm)		(6000 cpm)		(6000 cpm)
		(50 per cent)		(50 per cent)
(*ii*) With competition				
12Ag* + 6Ag + 6Ab	=	4 Ag* Ab	+	8 Ag*
(12000 cpm)		(4000 cpm)		(8000 cpm)
		+		+
		2 Ag Ab		4 Ag
		(33.33 per cent)		(66.67 per cent)

where,

Ag* = labelled antigen

Ag = unlabelled antigen (from standard or samples)

Ab = antibodies at a dilution which can bind 50 per cent of Ag in absence of Ag.

AgAb / Ag Ab = antibody complex with antigen (Ag or Ag)-Bound fraction

cpm = counts per minutes

Hence, as the concentration of unlabelled antigen (from standard/sample) increases, the competition for binding to Ab also increases, resulting in lowering of the per cent binding with labelled antigen (Ag*). These functions, namely, per cent bound (or per cent free) and concentration of Ag (standard) are plotted and concentration of unknown is thus read off the curve.

Exp. 5: Radioimmunoassay of T_3 and T_4

Reagents

(i) Diluent Buffer

Weigh 30 gm of glycine and 12 gm of NaOH in a beaker and add 500 ml DDW and adjust the pH to 8.7 with glacial acetic acid; about 15.6 ml of acetic acid is consumed. Make up the volume to 2 litre with DDW. It is stored at 4°C and should not be used after 4 weeks.

(ii) Normal Rabbit Serum (NRS)

The normal healthy rabbits are bled and serum separated. The serum is passed through Millipore. Aliquots of 2-3 ml are stored frozen at –20°C.

(iii) Assay Buffer

1 ml of assay buffer is required for each assay tube. The following is the composition of assay buffer:

1. 8 Anilino-1-Naphthalene-Sulfonic acid (ANS) 0.02 gm
2. Normal rabbit serum
   - (*a*) for charcoal separation method 1 ml or
   - (b) for double antibody separated method 0.25 ml (For T_3 only)

Make up the volume to 100 ml diluent buffer. Prepare the buffer fresh on the day of assay.

(iv) *Preparation of* ^{125}*I-labelled* T_3 *and* T_4

3,5-L-T_2 and 3,5,3′, L-T_3 (Sigma) are used as substrate for iodination of T_3 and T_4 respectively. Iodination is performed using a modified chloramine-T method to prepare high specific activity ^{125}I-labelled T_3 and T_4 (1). Each standard iodination is carried out with 1 µg (1 µl) of respective thyronies (1 mg/ml solution in 50 per cent aqueous propylene glycol, pH 10). To 500 uCi of ^{125}I in 5 µl volume is added which is followed by the addition of 20 µl of PO_4 (0.25 M, pH 7.4). The reaction is initiated by adding 50 µg of chloramine-T in 10 µl of 0.25 M PO_4 buffer pH 7.4. The reaction is allowed to proceed for 45 seconds and is terminated with the addition of 96 µg sodium metabisulphite in 20 µl of 0.25 M PO_4 buffer. Then, add immediately 50 µl of 50 per cent aqueous propylene glycol to the iodination mixture.

The iodination mixture is then very carefully loaded on a uniformly packed Sephadex-LH-20 column (25 c 1 cm) and is allowed to seep into the bed of the column and then start eluting with elution solvent which consists of ethylacetate: methanol: 2N ammonia (20:50:2) (The packed LH-20 column is equilibrated with the eluent before use). Generally 250 fractions of 40 drops each at an average flow rate of 20 drops/min. are collected preferably in a cold room (4-9°C). The fractions of interest are pooled and kept at –20°C in dark.

Working dilution of Tracer: The labelled T_3 or T_4 is diluted with diluent buffer to a counting rate of about 4000-6000 cpm.µl^{-1}.

(*v*) *Antiserum*

T_3-Antiserum: A very specific and high titre T_3-antiserum raised in rabbits is being used in a laboratory which is working at a final dilution of 1:10000. This antisera was raised against T_3-BSA conjugate.

T_4-Antiserum: T_4-BSA conjugate, coupled by carbodimide reagent, is mixed with complete Freund's Adjuvant (CFA) and injected in four rabbits at multiple sites (200 µg of conjugated/rabbit). The boost injections are given at an intereval of 4–6 weeks and the animals are bled 10 days after each booster injection and antibody titres are assayed.

(vi) Assay Buffer-Tracer Cocktail (B-T Cocktail)

For preparing B-T cocktail, 100 ml of Assay Buffer and 100 µl of Working Tracer solution (125 I-T or 125 I-T_4) are mixed thoroughly. Choice of the buffer is dependent on the method of separation for free and bound to be used (noted that higher concentration of NRS is used in Assay Buffer for charcoal separation method). This cocktail is prepared on the day of assay. Left over reagent is not stored.

(vi) Assay Buffer-Tracer-Antibody Cocktail (B-A-Ab Cocktail)

To 50 ml of B-T cocktail, add 0.25 ml of appropriately diluted antihormone antiserum (anti-T_3 and anti-T_4) and mix thoroughly. This cocktail is prepared just before the time it is required to be added in the assay tubes.

Both B-T cocktail and B-T-Ab cocktail are prepared in quantity just sufficient for the assay and unutilized reagents are discarded as they are not reused. These two reagents need thorough mixing specially because of the fact that the comparatively smaller volume of reagents like tracer or antibody are mixed with larger volume of buffer.

(vii) Standards

(*a*) *T_3 standards:* 3,5,3′-T_3 (Sigma) is used as standard for T_3 assay. Prepare a stock standard solution of T_3 (1 µg/µl) by first dissolving 1 mg T_3 in 50 µl of 0.1N NaOH and making up the final volume to 1 ml with freshly prepared 50 per cent aqueous propylene glycol. Small aliquots are stored at –20°C in dark.

Take 1 µl of the above solution and dilute to 10 ml of 50 per cent propylene glycol. Make aliquots of 200 µl each and store at –20°C in dark. This solution (Solution A) contains 100 µg Ta/µl.

50 µl of solution (A) is diluted to 200 µl with freshly prepared 50 per cent aqueous propylene glycol just prior to the addition in the assay tubes. This solution (Solution B) contains 25 µg T_3/µl.

(*b*) *T_4 standards:* 1 mg of 3,5,3′,5′-L-T_4 obtained from Sigma Chemicals is dissolved in 50 µl of 0.1 N NaOH and made up to 1 ml with 50 per cent aqueous propylene glycol to give a stock standard solution(A) of 1 mg/µl. The working standard solution (B) is made by adding 1 µl of 'A' in 10 ml of 50 per cent aqueous propylene glycol (100 pg/µl).

(viii) Hormone-Free Serum (HFS)

Pooled human serum is made free of T_3 and T_4 by first carefully adjusting its pH to 10 with 1N NaOH keeping it cool and stired all the time to avoid protein precipitation. The anion exchanger (QAE Sephadex A-25) is also brought to the same pH with 1 N NaOH and a column of 4 ml bed volume (20 0.8 cm) is prepared. 3 ml pooled serum is then passed through the column. The eluate is discareded. About 70 ml of the pooled serum is then added and the eluate collected and pooled. The above procedure is repeated again with a fresh QAE-Sephadex A-25 column. The pH of the serum is brought back to 7.5 with 0.1 N HCl. The serum is stored in aliquots at –20°C.

(ix) Second Antibody (Ab2)

The second antibody for separation of bound and free fractions is raised in goats against rabbit y-globulins. This antiserum is used at a final dilution of 1:20. Undiluted antisera are stored in aliquots at –29°C.

(x) Charcoal Suspension

1 per cent activated charcoal suspension is prepared in diluent buffer stored at 4°C. Use the suspension upto one month after the date of preparation.

Methodology

The procedures for T_3 and T_4 assay are exactly similar.

(a) Assay Protocol

The standards are assayed in triplicates. The first three tubes are for Blank (or non-specific binding, NSB), followed by three tubes of zero standard (Bo) and then from tube No. 7 onwards are the standards (8 sets for T_3 and 6 sets for T_4). For each sample take four tubes (two for sample blank and two for test sample). Protocol of T_3 and T_4 RIA has been given below.

The following components of assay are added in the sequence give below:

1. Standards and samples. Add standards from the working dilutions of standard solution using 5 and 10 µl fixed

needle Hamilton syringes, samples for T_4 were diluted 1/10 in Assay Buffer.

Add 0.1 ml sample (Diluted 1/10 in Assay Buffer containing 10 per cent HFS) in sample blank tubes and the corresponding test sample tubes.

2. Hormone free sera (HFS). 0.1 ml HFS (10 per cent in assay buffer for T_4) is added in Blank, Bo and Standards (Nos. 1-30 in the case of T_3 and 1-24 in the case of T_4 assay).
3. B-T cocktail: 0.9 ml of B-T cocktail (containing ^{125}I-T_3 or ^{125}I-T_4) is added in blank and all sample blank tubes.
4. B-T-Ab cocktail: 0.9 ml of this reagent (containing anti T_3 or anti T_4) is added in all assays tubes except blank (NSB) and Sample Blank tubes.

(b) Incubation

After mixing on vortex mixer, the tubes are kept at 4°C for 20-24 hrs incubation.

(c) Counting for 'Total cpm'

Every 10th tube of assay is counted in a gamma counter for 1 minute for accounting the total radioactivity added in the assay tubes.

(d) Separation of Bound and Free Fractions

Separation of bound and free fraction can be achieved by either charcola or double antibody method (The method to be followed should be decided at the start of assay as one has to use the appropriate assay buffer accordingly).

1. With charcoal: 100 µl of charcoal suspension is added in each tube. Mix the tubes on vortex mixer and centrifuge for 30-40 minutes at a speed of 3000 rpm at 4-8 °C. Decant the tubes immediately and keep inverted in beaker with a disc of an absorbent paper at the bottom. Wipe off the tubes carefully with wicks prepared by rolling tissue paper. The pellets are countered for one minute which give the measure of free fraction.
2. With second antibody: 50 µl of second antibody is added in each tube and mixed on a vortex mixer. The tubes are

kept at 4°C for an overnight incubation (16-20 hrs) and then centrifuged at a speed of 3000 rpm at 4°C for about 45 min. Decant the tubes immediately, keep them inverted and wipe carefully the drops of supernatant with the tissue paper wicks as described for charcoal separation method above. The pellets are counted for one minute in a gamma counter for the bound fraction.

Calculations

1. Calculate the mean of the replicate bound and free counts for each set of the assay as folows:
   (*a*) Charcoal Method
   ☆ Mean of the counts in pellet = Mean free counts
   ☆ Mean total countes-Mean counts in pellet=Mean bound counts
   (*b*) Doule Antibody Method
   ☆ Mean of the counts in pellet = Mean bound counts
   ☆ Mean total counts-Mean counts in pellet=Mean free counts
2. Next, calculate the ratio of 'Bound' and 'Free', counts (B/F) by dividing Mean Bound Counts by Mean Free Counts.
3. Calculate the Net B/F as follows:
   (*a*) Substract B/F of blank (NSB) from the B/F calculated for Bo and all the standards.
   (*b*) Substract B/F of sample blank from the B/F of respective sample (Test sample).
4. Using an ordinary graph paper (cm scale), plot B/F on Y axis and chromone/tube on the X-axis (dose response curve).

If agreed the value of unknwon (sample) for the corresponding. Net B/F of the sample from the curve; the value obtained is the concentrate. 20 n of hormone/0.1 ml sample (*i.e.* sample volume used in the assay). This value should be multiplied by 1000 to get the concentration/dl of sample.

A typical data obtained for dose reponse curve have been plotted in table.

Precautions

1. Care should be taken to maintain a temperature of about during the dispensing of reagents in the assay tubes as well as during incubation period or periods (double antibody method).
2. All reagents including antisera, NRS and HFS should have 0.01 per cent sodium azide as preservative.
3. Serum or plasma should be assayed as early as posible but can be stored at –20°C for a period upto 4 months without appreciable loss of hormones in the sample.

Exp. 6: Radioimmunoassay of Cortisol

Reagents

(i) Gelatin Phosphate Buffer (Assay Buffer)

Dissolve 1 gm gelatin in about 100 ml of warm distilled water with constant stirring. Dissolve the following salts in about 400 ml of distilled water in a separate beaker.

1. 2.35 g NaH_2PO_4 (MW 120)
2. 11.6 g Na_2HPO_4 (MW 142)
3. 8.8 g NaCl
4. 0.1 g NaCl

Mix the two solutions and make the volume to 1 litre with distilled water. The pH should be between 7.2-7.4. Store at 4°C. Prepare the fresh buffer after expiry of one month.

(ii) Cortisol Standard (Stock)

Prepare standard cortisol (Sigma) equivalent to 400 ng.ml in ethanol. Distribute 1 ml aliquot in screw cap sample tubes. Freeze dry the aliquots. Lyophillized as well as reconstituted (with assay buffer) aliquots are stored at 4°C. Reconstituted stock standards should not be used after 5 weeks from the date of reconstitution as cortisol has limited stability in aqueous medium.

(iii) Anti-cortisol Serum

Antiserum is raised in rabbit by immunization of animals with BSA-linked cortisol. 0.1 ml aliquots are kept at or below –20°C. The aliquot is appropriately diluted (generally more than 3000 times) with assay buffer to give 'Bo' between 45-55 per cent.

The antiserum we have been using, showed negligible cross-reaction with progesterone, corticosterone and 11–hydroxyprogesterone. It has shown more than 20 per cent cross-reaction with cortisone and about 10 per cent with 17–hydroxyprogesterone. However, it has shown a high cross-reaction (40 per cent) with compound S and hence it cannot be used for estimation of cortisol in children with congenital adrenal hyperplasia.

(iv) ^{5}H Cortisol solution (stock)

The tracer is purchased from Amersham International Limited, Amersham, England. The contents of ampoule (250 uCi) should be transferred carefully in a 25 ml volumetric flask and made up to the volume with a mixture of benzene and ethanol 9:1 (v/v) to give a solution containing 10 uCi of radioactivity/ml. Use only the high purity solvents. This solution remains stable for 6 months when stored at 4°C.

(v) Dextran-Charcoal Suspension

Dissolve 0.0625 g of dextran T-70 in 100 ml assay buffer and then add 0.625 g activated charcoal. Shake vigorously. Store at 4°C. This suspension can be used upto one month.

Methodology

(i) Preparation of Samples

Take 1 ml of distilled water in a test tube, add to it 50 µl of serum for plasma, cover it tightly and incubate for 30 minutes at 60°C in a water bath to denature the cortisol binding globulins.

(ii) Preparation of Working Dilutions

1. Standards: In a 10 ml capacity test tube; add 4.8 ml assay buffer and 200 µl cortisol standard from the reconstituted solution. Mix and mark the tube as A. This stock contains 16 g cortisol/ml. Take a set of 5 test tubes of same size and mark them serially from 1 to 5. Add 1 ml assay buffer in each. Add 1 ml solution from tube A to tube 1. Mix thoroughly and take 1 ml from tube 1 and add in tube 2. Mix and repeat the process upto tube 5. 100 µl solution of these dilutions will provide the following concentration of cortisol when added to assay tube.

Standard cortisol (pg/100 µl aliquot)	Tube No.
50	Tube 5
100	Tube 4
200	Tube 3
400	Tube 2
800	Tube 1
1600	Tube A

2 Tracer (^{3}H Cortisol). For each assay of 100 tubes, 150 µl of stock ^{3}H-cortisol solution is taken in 25 ml capacity test tube. The solvent is evaporated to dryness by passing slow stream of nitrogen gas. Dissolve the residue in 15 ml assay buffer, vortex mix and allow to stand in cold for about 15 minute before use. This solution should contain about 5000 cpm/0.1 ml aliquot.

(iii) Assay Protocol

The assay is carried out in 12 x75 mm glass tubes according to protocol described below. All the sets are in triplicate (tubes nos. are given in parentheses).

All the assay tubes and the reagents added are kept at 4°C.

1.. Blank (1-3): Add 600 µl assay buffer. Do not add antiserum in these tubes.
2. Bo tubes (4-6): Add 500 µl of assay buffer
3. Standards (7-24): 100 µl of standards (six dilutions) are added in triplicate in tube nos. 7-24. Add 400 µl of assay buffer in all standard tubes.
4. Samples: From tube No. 25 onwards add 100 µl of the diluted and incubated sera samples as described in Section B(1) above. Add 400 µl of assay buffer to all sample tubes.
5. Internal quality control: At least two internal quality control sera of low and high cortisol content should be used along with the samples.
6. Total count vials (T1, T2, T3): Add 100 µl aliquots of working dilution of tracer directly in 3 scintillation vials

containing 1 ml of double distilled alcohol. Add 800 µl of assay buffer also.

(iv) Addition of Labelled Cortisol and Antiserum

Add 100 µl of working dilution of labelled cortisol in all assay tubes. Add 100 µl of diluted antisera in all the tubes except in blank (1-3). Vortex mix all the assay tubes, cover them with aluminium foil and keep for incubation at 4°C for 20 to 24 hours.

(v) Separation of Bound and Free Fraction

Add 200 µl of dextran-charcoal suspension to each tube. The instructions given below should be strictly followed:

1. All the assay tubes and dextran-charcoal suspension should be kept at 4°C.
2. Charcoal suspension should not be allowed to settle and it should be kept constantly stirred.
3.. Time taken for adding in the first and the last tube should be kept to minimum.

Mix the tubes and leave them at 4°C for exactly 15 minutes. Spin the tubes for 10 minutes at 500g.

Quickly decant the supernatant directly into scintillation vials containing 1 ml double distilled alcohol. Add 10 ml scintillation fluid (4 g PPO and 0.5 g POPOP dissolved in 1 litre of toluence) to all the vials including total count vials and after allowing one hour for equilibration, count the radioactivity in a liquid scintillation counter for 5 minutes each.

Calculations

Logit-lot paper is used to plot dose responsive curve. The values of unknown determined from standard curve are divided by 50 to get the concentration of cortisol in µg/100 ml of serum or plasma.

The cortisol levels in normal human subjects is 5-18 µg/100 ml of serum or plasma. As the plasma cortisol undergoes diurnal rhythm,being high in the morning and low in the evening, the time of collection of blood samples is very important.

The data obtained for one of the typical dose response curve have been shown in table. Concentration of hormone in sample is calculated from the following formula:

$$X = (y - A)/B$$

where,

X = (log) concentration of unknown cortisol

Y = Logit (B/Bo)

A = Intercept at Y

B = Slope

For the above example, the intercept +S.E. is 5.31 + 0.08 and the slope + S.E. is –2.29 + 0.03. Antilog of the value obtained by the formula is divided by 50 to get the concentration of cortisol in µg/100 ml sample.

Conversion Factors

µmol/L = µg/ml 2.76

µg/ml = µmol/L 0.362

Data for Dose Response of Cortisol

Tube No.	*Assay Set*	*Bound cpm Replicates* A	b	c	*Mean*	*% B (Bound Blank)*	*(B/Bo) × 100*	*Logit (B/Bo)*
1-3	Blank	112	99	134	115	(1.94)		
4–6	Bo	2655	2581	2606	2614	42.11	100.0	
7–9	50	2131	2087	2165	2128	33.92	80.54	+1.42
10–12	100	1818	1831	1737	1796	28.32	67.25	+0.72
13–15	200	1373	1393	1463	1410	21.82	51.81	+0.07
16–18	400	966	959	955	960	14.24	33.81	–0.67
19–21	800	605	628	600	611	8.36	19.85	–1.40
22–24	1600	416	405	424	415	6.99	12.00	–1.99
T_1–T_2 Total counts		6004	5870		5928	5934		

C 10 per cent HS Before Addition

T_3 Assay			Tube Nos.	T_4 Assay	
Standard Solution A (µl)	*Standard Solution B (µl)*	*Conc Tube (pg)*		*Standard Solution B (µl)*	*Conc Tube (pg)*
–	1	25	7–9	1	100
–	2	50	10–12	2	200
–	3	75	13–15	4	400
1	–	100	16–18	6	600
1.5	–	150	19–21	8	800
2	–	200	22–24	10	1000
3	–	300	25–27	–	–
4	–	400	28–30	–	–

Appendix-I

pH values of various body material

Body fluids and tissues	7.35–7.45
Blood serum	7.35–7.45
Cerebrospinal fluid	7.4
Aqueous humour of eye	6.35–6.85
Saliva	about 0.9
Pure gastric juice	7.5–8.0
Pancreatic juice	7.0–8.0
Intestinal juice	7.4–8.5
Hepatic duct bile	5.4–6.9
Gallbladder bile	4.8–7.5
Urine	7.0–7.5
Faeces	7.4
Tears	6.6–6.9
Orange juice	2.6–4.4
Grapefruit juice	3.2
Tomatoes (ripe)	4.3
Egg white (fresh)	8.0

Appendix-II

Some Commonly Used Indicators of pH

Common Name pH	*PK in*	*Acid and Alkali Colours and Useful Range*
Thymol blue	1.7	Red-yellow (1.2-2.8)
Methyl orange	3.5	Red-yellow (3.1-4.4)
Bromocresol green	4.7	Yellow-blue (3.5.4)
Bromophenol blue	4.0	Yellow-blue (3.0-4.6)
Methyl red	5.1	Red-yellow (4.2-6.3)
Chlorphenol red	6.0	Yellow-red (5.1-6.7)
Bromocresol purple	6.2	Yellow-purple (5.4-7)
Bromothymol blue	7.0	Yellow-blue (6.0-7.6)
Phenol red	7.9	Yellow-red (6.8-8.4)
Cresol red	8.3	Yellow-red (7.2-8.8)
Thymol red	8.3	Yellow-blue (8.0)
Phenolphthalein	9.7	Colorless-red (8.3-10.0)
Thymolphthalein	9.9	Yellow-blue (9.3-10.5)
Neutral red	6.85	Yellow orange (6.8-8.0)

Exercise No. 3

Isolation and Identification of Virus by Different Methods

Take the tissue sample *i.e.* Kidney, Tail region, Head region, Liver etc.

↓

Homogenate the samples

↓

Centrifuge the samples at 3000 rpm at 4°C for 15 Min (VHS–4200)

↓

Take out the supernatant (150 μl) and add 150 μl antiserum of IPNV- SP for growth (For IHNV and VHS)

↓

Incubate the samples for one hour at -20°C

↓

Take out the samples from incubator and pour the media (L-15) 1ml and add 100 μl of sample in the wells 1& 3. Then take out the 100μl of media and sample from 1& 3 and then pour it into 2 & 4 well.

↓

Incubate it for 7 days at 14°C

3.1 For IPNV

Take the tissue sample *i.e.* Kidney, Tail region, Head region, Liver etc.

↓

Homogenate the samples

↓

Centrifuge the samples at 3000 rpm at 4°C for 15 min

↓

Take the pellets and add 3-4 ml E-MEM

↓

Sonicate the samples for 20–30 seconds (To open the cells)

↓

Centrifuge the samples at 4°C at 3000 rpm

↓

Take out 5 ml of sample and 5 ml media in flask containing CHSE and incubate it for 6 days at 22°C

3.2 Cell Culture

For IPNV: CHSE (Chinook Salmon Embryo Cells)

Take the flask of cell culture (250 cm^2) already containing cells.

↓

Wash with PBS (10 ml)

↓

Remove the supernatant

↓

Add 5 ml Trypsin EDTA for 5 min (Take care that cells start moving)

↓

Discard the supernatant

↓

Give a second wash of 5 ml Trypsin EDTA

↓

Neutralise the trypsin by adding 5 ml of media (EMEM)

↓

Add fresh media in different flask depending upon the size of flask (50 ml)

↓

Add the cell suspension in two flasks each (5 ml)

↓

Incubate the cells at 22°C for one week

↓

Take out the culture flasks and observe under microscope

↓

75 per cent cells should be grown atleast

↓

Add 50 µl of virus culture

↓

Allow it to grow at 15 °C in incubator for 7-10 days

↓

After 7 days observe under microscope. If virus grows well then store it in –80°C after transferring into vials.
Keep one vial for titration

↓

3.3 Titration of Virus

Take two 96 well plate in which CHSE cells are already grown.

↓

Infect the cells with virus 10 µl and take two lines as control and make a dilution 10^0, 10^1, 10^2, 10^3 - - - - -

↓

Incubate at 15°C for 7 days and calculate the $TCID_{50}$

↓

Read the 96 well plate under the microscope for CPE (Cytopathetic effect) +ve and

-ve and record it.

↓

Now calculate the $TCID_{50}$/ml using formula:

$$\frac{(\%\ \text{Positive Above } 50\%) - 50\%}{(\%\ \text{Positive Above } 50\%) - (\%\ \text{Positive Below } 50\%)} = \begin{array}{c}\text{Proportionate}\\ \text{Distance}\end{array}$$

(Log dilution above 50 per cent) + (Proportionate distance log dilution factor) = Log ID_{50}

↓

This is the end point of dilution *i.e.* the dilution that will infect 50 per cent of the test units inoculated

Example:

Log of Virus	*Infected Tests Units*	*Cumulative Infected (A)*	*Cumulative Non-infected*	*Per cent (%)*
10^{-1}	*6/6	56	0	100
10^{-2}	6/6	50**	0	100
10^{-3}	6/6	44	0	100
10^{-4}	6/6	38	0	100
10^{-5}	6/6	32	0	100
10^{-6}	6/6	26	0	100
10^{-7}	6/6	20	0	100
10^{-8}	6/6	14	0	100
10^{-9}	5/6	8	1***	87.50****
10^{-10}	3/6	3	4	33.33
10^{-11}	0/6	0	10	—
10^{-12}	0/6	0	16	—

Note: *Total number of all the infected units in 96 well plate and add from 10^{-1} 10^{-12}

** Total number of all the infected units in 96 well plate from 10^{-1} 10^{-12} (-) first row cells *i.e.* 10^{-1}

*** Total no. of non-infected cells in 96 well plate

**** Non-infected divided by cumulative multiply by 100 and minus 100.

$$\frac{87.50 - 50}{87.50 - 33.33} = \frac{37.50}{54.17} = 0.69$$

$(-9) + (0.69 \quad -1.0) = -9.69$

$TCID50 = 1x10^{-9.69}$

3.4 For IHNV: EPC (Epithelioma Papulosum of Carp)

Take the flask of 75 cm^2 already containing cells

↓

Wash with 6ml of PBS

↓

Add 3 ml Trypsin EDTA for 2 min (Take care that cells start moving)

↓

Discard the supernatant

↓

Give a second wash of 3 ml Trypsin EDTA

↓

Neutralise the trypsin by adding 3 ml of media (L-15, containing L- glutamine-1 per cent; FBS–5 per cent)

↓

Add fresh media and dispense and incubate at 22°C for 7 days

3.5 For VHSV: RTG2 (Rainbow Trout Gonad)

Take the flask of 75 cm^2 already containing cells

↓

Wash with 6ml of PBS

↓

Add 3 ml Trypsin EDTA for 2 min (Take care that cells start moving)

↓

Discard the supernatant

↓

Give a second wash of 3 ml Trypsin EDTA

↓

Neutralise the trypsin by adding 3 ml of media (L-15, containing L- glutamine-1 per cent; FBS–5 per cent)

↓

Add fresh media and dispense and incubate at 22°C for 7 days

3.6 RTG2-P1 Cell Line Culture

Take the TC flask in which RTG-P1 cells are already grown

↓

Remove the supernatant

↓

Wash with 10ml of PBS

↓

Remove the supernatant PBS

↓

Add 3 ml Trypsin EDTA for 2 min (Take care that cells start moving)

↓

Discard the supernatant

↓

Give a second wash of 3ml Trypsin EDTA

↓

Neutralise the trypsin by adding 3 ml of media (L-15, containing10ml FCB) and resuspend them with bulb vigorously to avoid clumps

↓

Take 30 ml of L-15 medium in each flask (5 Flasks)

↓

Add 2 ml of cell suspension in each flask containing already L-15

↓

Incubate at 22°C for 7 days

↓

Divide or multiply the cells after every 15 days following the same procedure

3.7 Split of RTG-P1 Cells into (12/24/48/96) Well Plate

Take out the flask (75 cm^2) from the incubator

↓

Remove the supernatant

↓

Wash with 10ml of PBS

↓

Remove the supernatant PBS

↓

Add 3 ml Trypsin EDTA for 2 min
(Take care that cells start moving)

↓

Discard the supernatant

↓

Give a second wash of 3ml Trypsin EDTA

↓

Neutralise the trypsin by adding 3 ml of media (L-15, containing10ml FCB) and resuspend them with bulb vigorously to avoid clumps

↓

Take 1ml of cell suspension in one bottle and add 20ml of medium into it

↓

Dispense the suspension into 20 wells

Calculation:

$$\frac{75 \times 3}{1.9} = 118.4$$

$$\frac{20}{118.4} = 0.17 \times 6 = 1.02\,\text{ml} = 1\,\text{ml}$$

where,

75 = Area of flask (cm^2)

3 = Ratio of split (1:3)

1.9 = Area of a well in 24 well plate

20 = Wells to be prepared

6 = Solution in the flask after trypsinization (Trypsin + cell + Medium)

Note: The area of one well in 12 well plate = 4.0 cm^2

The area of one well in 24 well plate = 1.9 cm^2

The area of one well in 96 well plate = 0.32 cm^2

3.8 Storage of Cells (RTG2- P1) in Liquid Nitrogen

Take the TC flask in which RTG-P1 cells are already grown

↓

Remove the supernatant

↓

Wash the cells with 6ml of PBS

↓

Remove the supernatant PBS

↓

Add 3 ml Trypsin EDTA for 2 min (Take care that cells start moving)

↓

Discard the supernatant

↓

Give a second wash of 3ml Trypsin EDTA

↓

Neutralise the trypsin by adding 6 ml of media (L-15, containing L- glutamine 10 ml) and resuspend them with bulb vigorously to avoid clumps

↓

Centrifuge at 1000 rpm at 4°C for 10 min.

↓

Discard the supernatant and add 6 ml FCS to the pellets

↓

Now take 8 ml of L-15 medium and 2 ml of DMSO and mix them well in separate bottle. (FREEZING MIXTURE)

↓

Take 9 cryovials and label them with cell line, date, medium, conc. of cells, P-, and add 5ml cell suspension in each vial

↓

Add.5ml of freezing mixture in each vial and store them at –80°C in MRFROSTER (Containing alcohol) for 4 days

↓

Take out the vials and store them in liquid nitrogen

↓

Keep one vial for further propagation of cells (to see the viability)

3.9 Enzyme-linked Immunoassay (ELISA) for IPNV

Take out the ELISA kit from freeze

↓

Take the microtitre plate/ well precoated with specific antibody to IPNV

↓

Keep first two wells blank (Don't add anything in these wells)

↓

Add 50 µl –ve control in two wells

↓

Add 50 µl +ve control in two wells

↓

Add 50 µl of sample in two wells (Each sample should be a multiple of two)

↓

Add 50 µl of buffer in all the wells except blanks

↓

Cover the plate/well with a lid and wrap it in aluminium foil and incubate it for 1 hr at 37°C

↓

Take out the plate, aspirate the contents of the wells and wash three times (2.30 min each) with washing buffer (1/20) solution, fill the wells to the upper edge.

↓

Tap the plate on a clean absorbent paper/ tissue paper to remove liquid after each wash

↓

Add 50 µl of Rabbit antibody (1/30) in all the wells except blanks

↓

Cover the plate and incubate it at 37°C for 1 hr

↓

Take out the plate, aspirate the contents of the wells and wash three times (2.30 min each) with washing buffer (1/20) solution, fill the wells to the upper edge.

↓

Tap the plate on a clean absorbent paper/ tissue paper to remove liquid after each wash

↓

Prepare Conjugate and add 50 µl into each well

↓

Cover the plate and incubate it at 37°C for 1 hr

↓

Take out the plate, aspirate the contents of the wells and wash three times (2.30 min each) with washing buffer (1/20) solution, fill the wells to the upper edge.

↓

Tap the plate on a clean absorbent paper/ tissue paper to remove liquid after each wash

↓

Add 100µl of substrate into all wells except blanks. Keep the substrate in dark, as it is light sensitive

↓

Cover the plate with lid and aluminium foil and incubate it at 37°C for 10 min. keep out of light.

↓

Stop the reaction by adding 100µl of stopping solution into the wells in the same order and in the same intervals in which the substrate was added

↓

Read the colour intensity of each well at 450 nm against blank

INTERPRETATION OF RESULTS (ELISA)

Positive if:

The mean Absorbance of tested sample is greater than 0.1 and at least twice greater than the mean Absorbance of control negative antigen

↓

ELISA titre of the antigen is the last sample dilution giving mean absorbance value greater than 0.1

3.10 Preparation of IPNV Sample for ELISA

Take the tissue sample *i.e.* Kidney, Tail region, Head region, Liver etc.

↓

Homogenate the samples

↓

Centrifuge the samples at 3000 rpm at 4°C for 15 Min

↓

Take the pellets and add 3-4 ml E-MEM

↓

Sonicate the samples for 20–30 seconds (To open the cells)

↓

Centrifuge the samples at 4°C at 3000 rpm

↓

Take out.5 ml of sample and 5 ml media in flask containing CHSE and incubate it for 6 days 22°C and now use the supernatant for ELISA

Protocol for Mx analysis

3.11 RNA Extraction from Liver Tissue

Collect ~ 50mg liver tissue into 500µl RNABee and store in –80 C freezer

↓

Sonicate from frozen for about 20 seconds, or until the sample is only slightly warm, then keep on ice

↓

Add 10 per cent v/v chloroform, gently shake and leave for 5 minutes on ice and then centrifuge at 14000 rpm for 15 minutes at 4 C

↓

Carefully remove the upper aqueous layer into a clean RNAse free tube, so that the white interphase is not disturbed

↓

Add an equal volume of isopropanol, gently mix then centrifuge at 14000 rpm for 15 minutes at 4 C

↓

Pour off the liquid and wash the white pellet in 1ml ice cold 75 per cent ethanol in DEPC water. For maximal washing try to get the pellet dislodged from the bottom of the tube, then centrifuge at 14000 rpm for 15 minutes at 4

↓

Repeat the washing step mentioned above

↓

Finally pour off the liquid and air dry for 5-10 minutes, then resuspend the pellet in the minimum volume of DEPC water required *i.e.* 50-200µl

↓

To calculate the RNA concentration, dilute the sample to 1 in 100 in DEPC water and measure the absorbance at 260/280nm

RT-PCR Steps for Obtaining Beta Actin and Mx Products

Dilute sample to 0.5µg/µl and add 2µl to 2µl of oligo dt (0.5µg/µl, 18T's), preheat thermal cycler to 70 C and heat sample for 10 minutes then place immediately on ice

↓

Add the following reaction mix for reverse transcription:

- ☆ 5µl 5x buffer
- ☆ 1µl 50 units/µl reverse transcription
- ☆ 1.25µl 10mM d NTP master mix
- ☆ 13.75µl DEPC water

(Final volume = 25µl)

Then heat the sample on the following cycle: 42 C for 60 minutes then at 70 C for 15 minutes then place on ice

↓

Prepare the PCR reaction mix as follows:

- ☆ 2.5µl 10x buffer
- ☆ 0.5µl 10mM d NTP master mix

☆ 0.75µl 50mM $MgCl_2$

☆ 1µl 10µM working strength forward beta actin primer (GCCGGGTTCGCTGGAGATGA)

☆ 1µl 10µM working strength reverse beta actin primer (GCGTGGGGCAGAGCGTAACC)

☆ 0.125µl 0.5 units/µl TAQ polymerase (diluted from 5µ/µl in DEPC water)

☆ 1.5µl cDNA from step 2 above (vortex before pipetting)

☆ 17.625µl DEPC water.

(Final volume = 25µl)

Then heat the sample on the following cycle: 94 C for 2 minutes then 30 cycles of 94 C for 45 seconds, 55 C for 45 seconds, 72 C for 20 seconds, finally 72 C for 2 minutes then place on ice

↓

Repeat as above using Mx primers and using the following cycle programme: 94 C for 2 minutes then 35 cycles of 94 C for 45 seconds, 55 C for 45 seconds, 72 C for 30 seconds, finally 72 C for 2 minutes then place on ice

↓

Pour 2 per cent agarose-M gel containing 5µl/100ml ethidium bromide *i.e.* 0.8g agarose-M, 40ml TBE 1 buffer and 2µl ethidium bromide

↓

To prepare the samples to run in the gel add 6µl cold DEPC water and 2µl gel loading solution to 4µl sample and mix in a tube, then pipette 10µl into each sample well

↓

Run the samples with 100bp ladder prepared by mixing 10µl cold DEPC water, 2µl gel loading solution and 2µl of ladder, then pipette 7ml to the wells

↓

Run the gel submerged in TBE 0.5 buffer for about 30- 60 minutes at 140 Volts then screen under UV

3.12 RT-PCR for VP2 Gene of IPNV

Sampling and RNA Extraction

Take out the supernatant and scrap off the cells from bottom of the wells/25 cm^2 flasks containing tissue culture cells

↓

Take 400 µl of sample and add 800 µl of Trizol

↓

Freeze them at –20°C for storage upto 1 month or start extracting RNA

↓

Labelled universal containers or bijoux will be required for waste, Trizol, isopropanol and 75 per cent ethanol while a glass bijoux is required for chloroform. Trizol and chloroform must be dispensed in the cabinet, as fumes released from these chemicals are toxic. Aliquot the appropriate amounts of reagents into the labelled Universal containers or bijoux.

↓

Add 800µl Trizol for blank sample in 400µl of DEPC water tube

↓

Keep it for 5 min

↓

Add 160 µl of chloroform in all the samples including blank and shake for 3 min

↓

Vortex for 15 sec

↓

Centrifuge at 13000 rpm for 15 min at 4°C

↓

Discard the chloroform (any) into waste bottle

↓

Take out the supernatant after centrifugation and add into another vial (Don't disturb the white layer)

↓

Add 500µl Isopropanol into each tube and keep them for 10 min. in the fume hood

↓

Centrifuge at 6500 rpm for 15 min at 4°C

↓

Remove the supernatant from all the samples

↓

Wash the pellets with 1ml of 75 per cent ethanol and centrifuge at 6500 rpm for 5 min at 4°C

↓

Discard the supernatant and air dry them for 3 min

↓

Resuspend the pellets in 15 µl DEPC water

↓

Add 147µl of DEPC water in different tube already labelled for corresponding sample

↓

Add 3µl from the samples into tubes already containing DEPC water (147µl)

Measurement of Conc. of RNA

Take 150µl of distilled water for blank sample in cuvette and read the absorbance

↓

Take 150µl of sample in cuvette and read the absorbance at 260 and 280nm wavelength and calculate the RNA conccentration

↓

Store the samples until use at –20°C

Calculation of Conc. of Sample

If 2 µg of RNA is required for reverse transcription, divide 2 by the concentration of the sample (*e.g.* 2/0.500 = 4.00). Subtract the amount of RNA required from 9.5, the volume required for reverse transcription (9.5–4.0 = 5.5) to determine the amount of dH_2O that is required to be added.

Preparation of RT and PCR Mix

Volumes of RT Reagents

Reagents	*x1 (µl)*	*x5 (µl)*	*x6 (µl)*	*x7 (µl)*	*x8 (µl)*	*x9 (µl)*	*x10 (µl)*
5 × 1st strand buffer	4	20	24	28	32	36	40
dNTPs (10mM of each)	2	10	12	14	16	18	20
Random hexamers (50ng/µl)	2	10	12	14	16	18	20
DTT (0.1M)	1	5	6	7	8	9	10
Rnase OUT (40U/µl)	0.5	2.5	3	3.5	4	4.5	5
MMLV-RT (200U/µl)	1	5	6	7	8	9	10

Add 9.5µl of sample or the amount calculated and incubate at 55-60°C for 10 min

↓

Add 10.5µl of RT mix into each sample

↓

Incubate at 37°C for 1 hr

PCR Reaction

Volumes of PCR Reagents

Reagents	*x1 (µl)*	*x5 (µl)*	*x6 (µl)*	*x7 (µl)*	*x8 (µl)*	*x9 (µl)*	*x10 (µl)*
10 × NH_4 buffer	5	25	30	35	40	45	50
$MgCl_2$ (1.5mM)	1.5	7.5	9	10.5	12	13.5	15
PRIMER-F	1	5	6	7	8	9	10
PRIMER-R	1	5	6	7	8	9	10
dNTP's	1	5	6	7	8	9	10

Prerpare the reaction mix according to table and number of samples

↓

Add PCR reaction mix (45µl) into small PCR tubes in clean room and add 2 drops of mineral oil on it

↓

Take out RT samples from incubator and add 5µl of sample (c-DNA) into each corresponding tube through oil

↓

Take the PCR tubes and switch on the Techne Genius thermal cycler and select the programme.

Programme

94 C–1 min

54 C–1 min } 35 cycles

72 C–1 min

72 C–5 mins

15 C–hold

↓

Start the programme, let the temperature rise to 90 C - 93 C then press the pause button. Put the PCR tubes in the PCR machine and leave for about 1 minute before releasing the pause button. The programme takes about 2 ½ hours.

↓

When the programme has finished, remove the PCR tubes from the thermal cycler and store in the fridge

Gel Preparation and Documentation

Principle

The PCR products are pipetted into wells in an agarose gel to which a current is applied. As DNA has a negative charge, it is attracted to the positive electrode. Smaller DNA fragments are able to pass through the gel faster than large fragments, and thus the

DNA fragments are separated by size. A DNA marker of known fragment sizes is run alongside the samples in order to determined the size of the resultant fragments.

Reagents

- ☆ Agarose–Invitrogen
- ☆ 10 TAE buffer
- ☆ 1 TAE buffer, prepared by adding 1 litre (+/- 10 per cent) of 10 TAE and 9 litres (+/- 10 per cent) of dH_2O with measuring cylinders to a labelled 10 litre carbuoy. Fill in a label for the carbouy, noting the reagent, the batch number, the date of preparation, and the date of expiry (one year from date of preparation) and the analyst's initial on the label. Record these details in the reagent record book located in the post PCR laboratory.
- ☆ Ethidium bromide (10mg/ml)–Sigma
- ☆ Distilled deionised water
- ☆ Gel Loading Solution–Sigma
- ☆ 50 µg/ml 50 bp ladder–Invitrogen

If an alternative supplier is used, the reagent must be of a similar chemical grade. Shelf lives of the reagents are as printed on the manufacturers label.

Equipment

- ☆ Paper towel
- ☆ Measuring cylinders
- ☆ Conical flask
- ☆ Loading tray geNunc module
- ☆ Calibrated P2, P10 & P20 pipettes
- ☆ Filter tips for P2, P10 & P20 pipettes
- ☆ Hoeffer HE 33 electrophoresis tank
- ☆ Hoeffer HE 33 gel casting tray
- ☆ Hoeffer HE 33 gel running tray
- ☆ Hoeffer HE 33 gel comb
- ☆ Life Technologies 250 Ex power pack
- ☆ Pharmacia electrophoresis power pack

- ☆ Mettler Balance 2 decimal places, calibrated and checked before use
- ☆ Either Syngene Gene Genius Gel Doc system
- ☆ or Polaroid camera and Polaroid 665 film
- ☆ Sharps container
- ☆ Biohazard bag/Stand.

Exercise No. 4

Recent Diagnostic Methods for Isolation and Identification of Infectious Pancreatic Necrosis Virus (IPNV)

The intensification and diversification of freshwater aquaculture practices in recent years have led to increased incidence of fish diseases in the country. A similar picture is prevalent with regard to brackish water and marine culture system. Fisheries suffer from a wide variety of diseases of infectious and non-infectious etiology in their natural environment. Abundant literature is available with regard to fish mortalities due to infection by viruses, bacteria, parasites, fungi, nutritional deficiencies, metabolic disorders and faulty environmental conditions like oxygen deficiency, high or very low water temperature, chemical pollutants, over-crowding, malnutrition, algal blooms, etc. However, fish mortalities are not the only criterion to evaluate the effect of fish diseases. Even the morbidity which leads to weight losses and poor growth in surviving fish contribute substantial losses to the farmers. The need for maintaining healthy stock as a prerequisite to enhance fish production levels has been fully appreciated by planners and farmers alike in the recent years.

4.1 Infectious Pancreatic Necrosis (IPN)

It is a highly contagious viral disease of young fish of the salmonid species. It is widely reported in Europe, Asia, America and Canada. Significant mortality can occur from IPN outbreaks in fry. Disease signs include: pale liver and spleen, stomach and intestine devoid of food, darker skin and trailing faecal casts. Infectious pancreatic necrosis virus (IPNV) can cause an asymptomatic carrier state in salmonid fish. These carriers can shed the virus and transmit the virus to their progeny and other susceptible species. It is, therefore important to detect asymptomatic carriers in order to limit the spread of infection.

Two new methods *i.e.* destructive and non-destructive were developed for the detection of IPNV carriers in Atlantic salmon (*Salmon salar* L.) and halibut (*Hippoglossus hippoglossus* L.). These methods are easy, inexpensive and very sensitive as compared to OIE (Office International des Epizootiesm 2000) standard tissue culture and RT-PCR. No information is available on the long-term persistence of IPNV in halibut and the fish are detected carriers by these methods. The destructive methods of viral isolation involve the need to kill the fish to obtain kidney and liver tissue. As halibut are repeat spawners and for broodstock testing there was a great need to develop a sensitive non-destructive method for isolation of IPNV. Macrophages are phagocytic cells and capable of killing wide range of pathogens. These play a vital role in immune response of fish and hence utilized in developing these methods.

4.2 Destructive Method

The macrophages are isolated following Secombes (1990) with some modifications. Head kidney (HK) from Atlantic salmon and halibut is dissected using sterile conditions into ice chilled 3ml cell separation medium. This medium consists of Leibovitz (L-15, PAA Laboratories) medium, supplemented with heparin 10U/ml, 2 per cent FCS, 50U/ml penicillin and 50µg/ml streptomycin (P/S). The HK is passed through a 100µm nylon mesh and left on ice for 10 min. The cell suspension is carefully layered onto a 34/51 per cent discontinuous Percoll density gradient. The gradients are centrifuged at 540g for 35 min at 4°C with low brake. The band of cells at the gradient interface is collected and diluted 3-4 times with low serum medium (L-15 medium containing 0.1 per cent FCS, P/S) and centrifuged at 540g for 15 min at 4°C. The supernatant is discarded and the pellets are resuspended in the low serum medium to eliminate the Percoll gradient and then centrifuged again as before. The supernatant is discarded and the cells are resuspended in1ml L-15 containing low serum medium. The cells are counted using a haemocytometer, their viability is assessed by trypan blue exclusion and adjusted to approximately 1x 10^7 cells/ml in low serum medium. Aliquots of 100µl are added to each well of a 96-well tissue culture plate (Corning Incorporated) and cells are left to attach overnight at 22°C. The Percoll is very costly and this method is time consuming and not feasible for large number of samples. So we have discarded the percoll gradient step and isolated the adherent macrophages from head kidney. The HK material (1 gram) passed through a 100µm nylon mesh using 2 ml of cell separation medium and 100µl of the cell suspension is added directly into each well of a 96-well tissue culture plate and cells are left to attach overnight at 22°C. To check whether this method is applicable for field samples, head kidney material is divided into 4 sections, transferred into cell separation medium and stored in an ice-packed polystyrene box. Macrophages are prepared after 0, 24, 48 and 72 hours of sampling. The non-percoll macrophages are compared with percoll macrophage for isolation of IPNV and similar results are found. This method is also compared with OIE standard tissue culture and RT-PCR and found more sensitive as compared to both.

4.3 Non-destructive Method

Blood is collected aseptically from the caudal vein of anaesthetised fish in heparinised vacuatte tubes (Greiner bio-one). The blood (0.5ml) is diluted three times with cell separation medium. The cell suspensions are layered onto 34/51 per cent Percoll discontinuous gradients. The gradients are then centrifuged at 540g for 35 min at 4°C with low brake. The band of cells at the gradient interface are collected and diluted 3-4 times with low serum medium and centrifuged using the same conditions as before. The supernatant is discarded and the pellets are resuspended in 1ml low serum medium. The cells are counted using a haemocytometer, their viability assessed by trypan blue exclusion and adjusted to approximately 1x 10^7 cells/ml in low serum medium. Aliquots of 100µl are added to each well of a 96-well tissue culture plate and cells are left to attach overnight at 22°C.

Whole blood or blood diluted three times with cell separation medium is aliquoted (100µl/well) into a 96-well tissue culture plate and cells are left to attach overnight at 22°C. In order to check whether this method is feasible for field samples, the blood is kept in heparinised vacuatte tubes and stored in an ice-packed polystyrene box. Blood leucocytes are prepared after 0, 24, 48 and 72 hours of sampling and compared with earlier methods. Sometime the fish is very small or not able to take out enough quantity of blood, different quantities (10, 50, 100, 250, 500µl) of blood is diluted with cell separation medium and aliquoted (100µl/well) into 96 well plate, incubated overnight and compared with earlier methods.

4.4 Inoculation of Samples into CHSE-214

After overnight incubation of kidney and blood leucocytes, the non-adherent cells are gently resuspended using a pipette and the supernatant containing the non-adherent cells from two wells is inoculated into CHSE-214 25cm^2 flasks. The adherent macrophages and blood leucocytes are washed 2-3 times with L-15 medium supplemented with 5 per cent FCS and P/S. After one hour, two wells of each macrophages and blood leucocytes are lysed for 15 min with 100µl Trypsin-EDTA lysis buffer (Invitrogen) per well. Any remaining adherent material is scraped off the bottom of the well and the well contents are passed into CHSE-214 25cm^2 flasks. Cultures are incubated at 15°C for seven days. The sample and control flasks are examined for the presence of CPE on days 3,7 after incubation and positive confirmation of IPNV infection is obtained using an IPNV-ELISA. The strain IPNV SP Spjarup (EU Reference Laboratory) is used for virus positive controls during each set of tests. Negative control CHSE-214 flasks are separately inoculated with 100µl of each of the reagents used in the processing of macrophages (cell separation medium, low serum medium, high serum medium, 34/51 per cent Percoll or trypsin/EDTA lysis buffer).

4.5 Macrophage Culture

Preparation of Solutions

L-15 Medium with 2 per cent FCS

500 ml L-15 without glutamine

10 ml FCS (2 per cent)

5 ml P/S(Penicillin and Streptomycin)

0.5ml Heparin (10 unit/ml)

L-15 Medium with 0.1 per cent FCS

500 ml L-15 without glutamine

0.5 ml FCS (0.1 per cent)

5 ml P/S(Penicillin and Streptomycin)

L-15 Medium with 5 per cent FCS

500 ml L-15 without glutamine

25 ml FCS (5 per cent)

5 ml P/S (Penicillin and Streptomycin)

Percoll Solutions

51 per cent Percoll

51 ml Percoll

10 ml x10HBSS

35 ml DW(deionized) first

add - - - N Hcl to adjust pH 7.2-7.4

add more water to make 100ml

34 per cent Percoll

34 ml Percoll

10 ml x10HBSS

50 ml DW(deionized) first

Add —N Hcl to adjust pH 7.2-7.4

Add more water to make 100ml

Procedure

Using sterile technique, dissect out the head kidney (HK) from a 500-700g fish

↓

Take out the HK and cut into 3-4 pieces and pass through 100 µl nylon mesh with L-15 medium containing 2 per cent FCS, 100 µl/ml P/S and 10µ/ml heparin

↓

Take 5 ml of 51 per cent percoll in 6 vials of 15ml and add 3 ml of 34 per cent percoll slowly not to disturb the other layer (Can be prepared in advance)

↓

Add 1 ml each of cell suspension in all the vials

↓

Centrifuge at 1600rpm for 25-30 Min at 4°C

↓

Discard the upper layer of RBC's

↓

Take a band of cells in between percoll layers

↓

Add 3-4 times L-15 supplemented with 0.1 per cent FCS and 100 µl/ml P/S and centrifuge at 1600rpm for 15-20 min at 4 °C to get rid of percoll

↓

Discard the supernatant and add 3-4 times L-15 and centrifuge at 1600rpm for 15-20 min at 4°C to get rid of percoll

↓

Discard the supernatant and add 1ml of L-15 to the pellets and resuspend it into all the 6 vials (Only 1ml of L-15 is to be added in all the vials)

↓

Take 20-30 µl and dilute it 25 times and add 20 µl of trypan blue

Count the cells in Haemocytometer under microscope

↓

Calculate the dilution factor by using the formula given below and add the required vol. of L-15 and adjust the cell suspension to 1 10^7 or 2 10^7 viable cells/ml

↓

Distribute it into 100µl of this suspension in 96 well plate

Formula

$$\frac{\text{No. of Cells} \times \text{No. of Squares} \times 10^4 \times \text{Dilution Factor}}{1 \times 10^7 \text{ or } 2 \times 10^7}$$

$$\frac{40\ (e.g.) \times 5 \times 10000 \times 25\,(e.g.)}{10^7} = \frac{50000000}{10000000} = 5$$

i.e. So you have to add 5 ml in one ml cell suspension and distribute it into 60 wells (100µl each)

↓

Infect the one day old macrophages (96 well plate) with IPNV Cole deep virus (851/99, Batch 2) from 10^{10} to 10^1. (If the medium is 100 µl then add 10µl; if 150 µl then add 15µl; and start from 10^{10} to 10^1). Keep three cells as control and make a replication of three wells for infection

↓

Incubate the cells at room temperature

↓

Take out the supernatant in Cryovials next day

↓

Add 500µl in CHSE flasks (cells should be 50 per cent grown) containing 5 ml media (1:10)

↓

Incubate the cells at 15°C for 7 days

↓

Now add 10-20µl of Trypsin EDTA in each well for 2-3 min. The cells starts dissolving in EDTA

↓

Add 30-40 µl of L-15 medium

↓

Take out the supernatant in bijoux tubes and sonicate the sample @75 per cent for 1 min

↓

Now add this supernatant in CHSE flasks containing 5ml cell suspension

↓

Incubate the samples at 15°C for 7 days

4.6 Enzyme-linked Immunosorbent Assay (ELISA) for IPNV

The method used is a double antibody sandwich ELISA using Rabbit anti-IPNV polyclonal antiserum to capture IPNV in the tissue culture supernatants and CE4 monoclonal antibodies (MAb) specific to IPNV followed by a goat anti-mouse alkaline phosphatase conjugate are used for IPNV detection. The protocol has been described in detail by Smail *et al.* (2003).

4.7 Titration of Virus from Macrophages

Following overnight incubation of kidney cell suspensions, with no prior fractionation on Percoll, in 96 well plates, the supernatant containing non-adherent cells is discarded and the adherent macrophages are washed and cultured in L-15, 5 per cent FCS, P/S. After 1 h in culture, the supernatants are discarded and the macrophages are lysed in Trypsin/EDTA buffer and serial dilutions of the macrophage lysates are inoculated into a 96-well CHSE –214 cell culture plate and observed for the development of CPE. The $TCID_{50}$/100µl is calculated seven days after inoculation.

4.8 Reverse Transcription Polymerase Chain Reaction (RT-PCR) for VP2 Gene for Isolation of IPNV

The reverse transcription-polymerase chain reaction for detection of IPNV is performed following Taksdal *et al.* (2001) with minor modifications on Atlantic salmon and halibut samples. RNA is extracted from whole kidney samples using TRIzol® reagent (Invitrogen) according to the manufacturer's recommendations. RNA is resuspended in DEPC-treated ddH_2O. Concentration and purity is quantified by measuring absorbance at 260 and 280nm (GeneQuant II, Pharmacia Biotech). In the RT step 2µg RNA is denatured for 10min at 55-60°C followed by cooling on ice. The reverse transcriptase mix (10.5µl) is added, giving a final volume of 20µl containing 1x 1st strand buffer (Invitrogen), 10mM of each deoxynucleoside triphosphate (Invitrogen), 2.5mM of random primer

(Promega, Inc), 0.1M DTT (Invitrogen), 20U RNaseOUT™ (Invitrogen), and 200U Moloney Murine Leukemia virus reverse transcriptase (MMLV-RT) (Invitrogen). This mixture is incubated for 1 hour at 37°C.

Five µl cDNA are added to 45µl of PCR mix, giving final concentrations of 1x Reaction Buffer (Bioline, London, UK), 1.5mM $MgCl_2$, 0.2mM deoxynucleoside triphosphates (each), 28pmol of each primer, 2 U of *Taq* polymerase (Bioline) and DEPC-treated ddH_2O. The reactions are overlaid with mineral oil (Sigma). The primers used for IPN detection are IPNV 1(forward, 5'-ATCTGCGGTGTAGACATCAAAG -3') and IPNV 2(reverse, 5'-TGCAGTTCCTCGTCCATCCC-3'). Amplification reactions are subjected to 35 cycles of 94°C for 1min, 60°C for 1min and 72°C for 1min followed by an extension step of 5min at 72°C. The PCR products are subjected to electrophoresis on a 1.5 per cent (w/v) gel in TAE buffer (40mM Tris, 20mM acetic acid, 2mM EDTA), stained with ethidium bromide and photographed using the Syngene GeneGenius Gel Documentation and Analysis system. (GeneSnap software version 4.0).

Exercise No. 5

Introduction to Freshwater Prawn (Scampi) *Macrobrachium rosenbergii* Diseases

It is believed that freshwater prawn (scampi), *Macrobrachium rosenbergii* is less susceptible to diseases than penaeid shrimp (*Panaeus monodon*). It might be as a result of lower stocking densities used in culture practices and less transfer of brood stock. However, a variety of diseases have been reported in larval, juvenile and adult scampi, which include fouling protozoans such as *Zoothamnium*, *Vorticella* and *Epistylis*; fungal pathogens such as *Lagenidium*, *Sirolpidium* and *Fusarium*; bacteria such as *Vibrio*, *Aeromonas*, *Leucothrix* etc., and also non-pathogenic diseases such as stress and neurotic symptoms, moult failure and so on. This may be due to shorter life cycle and faster metabolism of these animals, which results in weaker resistance to pathogens. Prawns are highly sensitive to pathogens during molting period when they cannot feed, have a week body and poor mobility. The incidence of disease is higher in low quality water, particularly if oxygen levels are low. Prawns need good quality water and enough plant life. The following diseases of freshwater prawn have been reported by various workers:

5.1 Larval Midcycle Disease

It widely occurs in Thailand, Philippines, Malaysia and India. The larval stages IV to XI are usually found to be affected from this disease. The symptoms appear about 10 days post-hatching and high larval mortality is observed. The weak larvae stop feeding and settle at the bottom of the rearing tank and exhibit spiral swimming behaviour along with vertical axis turning. The larvae become bluish grey in colour and the hepatopancreatic epithelium gets atrophied.

Possible Causes

This disease might be due to nutritional deficiency, since lecithin deficiency is often said to occur along with this disease. Bacteria, *Enterobacter aerogens* has been reported to be associated with this disease.

Preventive Measures

Infected stock should be immediately destroyed and tanks should be disinfected and dried. The farmers should adopt standard hatchery techniques that may prevent vertical or horizontal transmission of this disease.

5.2 Bacterial Diseases

Bacterial Necrosis

This is a common disease observed in larvae, post-larvae or adults in freshwater prawns. It is variously termed as 'black spot', 'brown spot', 'burnt spot', 'Shell Disease' or 'Chitinolytic bacterial disease'. This is caused by the invasion of chitinolytic bacteria, which break down the chitin of the exoskeleton, leading to erosion and melanization (dark brown to black pigmentation) at the site of infection. Several chitinolytic bacteria (Gram -ve rods) such as *Vibrio* spp., *Pseudomonas* spp., *Aeromonas* spp., *Spirillium* spp. etc are involved. This disease reduces the value of the harvested prawns, apart from causing mortalities.

This disease can be controlled in captive and cultured populations by reducing over crowding, proper husbandry and system hygiene.

Filamentous Bacterial Disease

Filamentous bacteria such as *Leucothrix muco, Thriothrix* spp., *Flexibacter* spp. etc. sometimes cause mortalities in freshwater prawns by discolouration of gills and associated secondary infections. The larvae or the post larvae become moribund, with reduced mortality, poor feeding and growth.

Increasing husbandry and hygiene standards will improve the situation.

Luminescent Bacterial Disease

Luminescent bacteria especially *Vibrio harvey* often infect the prawn larvae reared in hatchery conditions, when they appear in seawater. They spread very fast and poor hatchery conditions increase their virulence. The infected larvae become fluorescent in dark light, with reduced feeding and in severe cases mass mortalities occur.

Treatment

This disease can be cured by using specific antibiotics and disinfecting the pond water.

5.3 Fungus Disease

Larval mycosis due to *Fusarium* spp., *Lagendium* spp., *Sirolpidium* spp. cause severe mortalities to prawn larvae often as a secondary infection. *F. solani, F. moniliformes, L. callinectes* etc. are commonly found. Fungi infect the dead or damaged tissue caused by wounds or other infections resulting in locomotory difficulties due to mycelial growth. In serious infections of *Lagendium*, extensive non-septate highly branched mycelia invade through out the body replacing all the tissues. Specialized hyphae protrude through the cuticle.

5.4 Viral Disease

Scampi is believed to be free of viral diseases; however, a case of occurrence of white spot virus has been reported from Taiwan in *M. rosenbergii*. Signs closely resembling white spot syndrome have been observed on its exoskeleton as reported by Peng *et al*. (1998).

5.5 Protozoan Parasites

Fouling protozoans such as *Zoothamnium, Vorticella* etc. are of common occurrence in larval, post larval or adult scampi. They attach to the body and appendages and disrupt mobility and feeding. Mortality occurs only in severe cases of susceptibility. Formalin treatment will be helpful to save the affected individuals.

5.6 Other Diseases

White Muscle Disease

Opacity or 'white muscle' is a condition to the prawn arising out of any stress such as over crowding, extreme temperature or pH, low dissolved oxygen etc. The body appears opaque white proceeding forward from the tail region. The condition is often reversible once the stressed condition is corrected.

Blackening of Gills

Blackening of gills of scampi may sometimes be found as a manifestation of several other disease syndromes, precipitation of dissolved chemicals, turbidity, vitamin C deficiency etc. General discolouration of gills may occur due to melanization of tissue and necrosis, which may be visible through the side of carapace.

Index

L

M

N

O

P

Q

R